49p

James Benjamin Blish was born in America in 1921. He studied zoology at Rutgers University and served for two years in the US Army. During the last few years of his life, he moved with his family to England, where he died in 1975. His prolific output included book adaptations of the original Star Trek TV scripts and many highly acclaimed novels, amongst them *A Case of Conscience,* for which he won a Hugo Award in 1959.

Also by James Blish

James Blish

Galactic Cluster

A PANTHER BOOK

GRANADA
London Toronto Sydney New York

Published by Granada Publishing Limited in 1980

ISBN 0 586 04573 2

First published in Great Britain by
Faber and Faber Ltd 1960
Copyright © James Blish 1960

Granada Publishing Limited
Frogmore, St Albans, Herts AL2 2NF
and
3 Upper James Street, London W1R 4BP
866 United Nations Plaza, New York, NY 10017, USA
117 York Street, Sydney, NSW 2000, Australia
100 Skyway Avenue, Rexdale, Ontario, M9W 3A6, Canada
PO Box 84165, Greenside, 2034 Johannesburg, South Africa
61 Beach Road, Auckland, New Zealand

Set, printed and bound in Great Britain by
Cox & Wyman Ltd, Reading
Set in Intertype Times

Granada ®
Granada Publishing ®

To

KENNETH S. WHITE

Contents

Common Time

'. . . the days went slowly round and round, endless and un-
eventful as cycles in space. Time, and time-pieces! How
many centuries did my hammock tell, as pendulum-like it
swung to the ship's dull roll, and ticked the hours and ages.'

Herman Melville,
in *Mardi*

I

Don't move.

It was the first thought that came into Garrard's mind
when he awoke, and perhaps it saved his life. He lay where
he was, strapped against the padding, listening to the round
hum of the engines. That in itself was wrong; he should be
unable to hear the overdrive at all.

He thought to himself: *Has it begun already?*

Otherwise everything seemed normal. The DFC-3 had
crossed over into interstellar velocity, and he was still alive,
and the ship was still functioning. The ship should at this
moment be travelling at 22.4 times the speed of light – a neat
4,157,000 miles per second.

Somehow Garrard did not doubt that it was. On both
previous tries, the ships had whiffed away towards Alpha
Centauri at the proper moment when the overdrive should
have cut in; and the split-second of residual image after they
had vanished, subjected to spectroscopy, showed a Doppler
shift which tallied with the acceleration predicted for that
moment by Haertel.

The trouble was not that Brown and Cellini hadn't got
away in good order. It was simply that neither of them had
ever been heard from again.

Very slowly, he opened his eyes. His eyelids felt terrifically heavy. As far as he could judge from the pressure of the couch against his skin, the gravity was normal; nevertheless, moving his eyelids seemed almost an impossible job.

After long concentration, he got them fully open. The instrument-chassis was directly before him, extended over his diaphragm on its elbow-joint. Still without moving anything but his eyes – and those only with the utmost patience – he checked each of the meters. Velocity: 22.4 c. Operating temperature: normal. Ship temperature: 37° C. Air-pressure: 778 mm. Fuel: No. 1 tank full, No. 2 tank full, No. 3 tank full, No. 4 tank nine-tenths full. Gravity: 1 g. Calendar: stopped.

He looked at it closely, though his eyes seemed to focus very slowly, too. It was, of course, something more than a calendar – it was an all-purpose clock designed to show him the passage of seconds, as well as of the ten months his trip was supposed to take to the double star. But there was no doubt about it: the second-hand was motionless.

That was the second abnormality. Garrard felt an impulse to get up and see if he could start the clock again. Perhaps the trouble had been temporary and safely in the past. Immediately there sounded in his head the injunction he had drilled into himself for a full month before the trip had begun—

Don't move!

Don't move until you know the situation as far as it can be known without moving. Whatever it was that had snatched Brown and Cellini irretrievably beyond human ken was potent, and totally beyond anticipation. They had both been excellent men, intelligent, resourceful, trained to the point of diminishing returns and not a micron beyond that point – the best men in the Project. Preparations for every knowable kind of trouble had been built into their ships, as they had been built into the DFC–3. Therefore, if there was something wrong, nevertheless, it would be some-

thing that might strike from some commonplace quarter —
and strike only once.

He listened to the humming. It was even and placid, and
not very loud, but it disturbed him deeply. The overdrive
was supposed to be inaudible, and the tapes from the first
unmanned test-vehicles had recorded no such hum. The
noise did not appear to interfere with the overdrive's operation, or to indicate any failure in it. It was just an irrele-
vancy for which he could find no reason.

But the reason existed. Garrard did not intend to do so
much as draw another breath until he found out what it was.

Incredibly, he realized for the first time that he had not in
fact drawn one single breath since he had first come to.
Though he felt not the slightest discomfort, the discovery
called up so overwhelming a flash of panic that he very
nearly sat bolt upright on the couch. Luckily — or so it
seemed, after the panic had begun to ebb — the curious leth-
argy which had affected his eyelids appeared to involve his
whole body, for the impulse was gone before he could sum-
mon the energy to answer it. And the panic, poignant though
it had been for an instant, turned out to be wholly intellectual.
In a moment, he was observing that his failure to breathe in
no way discommoded him as far as he could tell — it was just
there, waiting to be explained—

Or to kill him. But it hadn't, yet.

Engines humming; eyelids heavy; breathing absent; calendar stopped. The four facts added up to nothing. The
temptation to move something — even if it were only a big
toe — was strong, but Garrard fought it back. He had been
awake only a short while — half an hour at most — and
already had noticed four abnormalities. There were bound
to be more, anomalies more subtle than these four; but
available to close examination before he had to move. Nor
was there anything in particular that he had to do, aside
from caring for his own wants; the Project, on the chance
that Brown's and Cellini's failure to return had resulted

from some tampering with the overdrive, had made everything in the DFC–3 subject only to the computer. In a very
real sense, Garrard was just along for the ride. Only when
the overdrive was off could he adjust—

Pock.

It was a soft, low-pitched noise, rather like a cork coming
out of a wine bottle. It seemed to have come just from the
right of the control-chassis. He halted a sudden jerk of his
head on the cushions towards it with a flat fiat of will.
Slowly, he moved his eyes in that direction.

He could see nothing that might have caused the sound.
The ship's temperature dial showed no change, which ruled
out a heat-noise from differential contraction or expansion –
the only possible explanation he could bring to mind.

He closed his eyes – a process which turned out to be just
as difficult as opening them had been – and tried to visualize
what the calendar had looked like when he had first come
out of anaesthesia. After he got a clear and – he was almost
sure – accurate picture, Garrard opened his eyes again.

The sound had been the calendar, advancing one second.
It was now motionless again, apparently stopped.

He did not know how long it took the second hand to
make that jump, normally; the question had never come up.
Certainly the jump, when it came at the end of each second,
had been too fast for the eye to follow.

Belatedly, he realized what all this cogitation was costing
him in terms of essential information. The calendar had
moved. Above all and before anything else, he *must* know
exactly how long it took it to move again—

He began to count, allowing an arbitrary five seconds lost.
One-and-a-six, one-and-a-seven, one-and-an-eight—

Garrard had got only that far when he found himself
plunged into Hell.

First, and utterly without reason, a sickening fear flooded
swiftly through his veins, becoming more and more intense.
His bowels began to knot, with infinite slowness. His whole

body became a field of small, slow pulses – not so much shaking him as putting his limbs into contrary joggling motions, and making his skin ripple gently under his clothing. Against the hum another sound became audible, a nearly subsonic thunder which seemed to be inside his head. Still the fear mounted, and with it came the pain, and the tenesmus – a board-like stiffening of his muscles, particularly across his abdomen and his shoulders, but affecting his forearms almost as grievously. He felt himself beginning, very gradually, to double at the middle, a motion about which he could do precisely nothing – a terrifying kind of dynamic paralysis ...

It lasted for hours. At the height of it, Garrard's mind, even his very personality, was washed out utterly; he was only a vessel of horror. When some few trickles of reason began to return over that burning desert of reasonless emotion, he found that he was sitting up on the cushions, and that with one arm he had thrust the control-chassis back on its elbow so that it no longer jutted over his body. His clothing was wet with perspiration, which stubbornly refused to evaporate or to cool him. And his lungs ached a little, although he could still detect no breathing.

What under God had happened? Was it this that had killed Brown and Cellini? For it would kill Garrard, too – of that he was sure, if it happened often. It would kill him even if it happened only twice more, if the next two things followed the first one closely. At the very best it would make a slobbering idiot of him; and though the computer might bring Garrard and the ship back to Earth, it would not be able to tell the Project about this tornado of senseless fear.

The calendar said that the eternity in hell had taken three seconds. As he looked at it in academic indignation, it said *Pock* and condescended to make the total seizure four seconds long. With grim determination, Garrard began to count again.

He took care to establish the counting as an absolutely

even, automatic process which would not stop at the back of his mind no matter what other problem he tackled along with it, or what emotional typhoons should interrupt him. Really compulsive counting cannot be stopped by anything – not the transports of love nor the agonies of empires. Garrard knew the dangers in deliberately setting up such a mechanism in his mind, but he also knew how desperately he needed to time that clock-tick. He was beginning to understand what had happened to him – but he needed exact measurement before he could put that understanding to use.

Of course there had been plenty of speculation on the possible effect of the overdrive on the subjective time of the pilot, but none of it had come to much. At any speed below the velocity of light, subjective and objective time were exactly the same as far as the pilot was concerned. For an observer on Earth, time aboard the ship would appear to be vastly slowed at near-light speeds; but for the pilot himself there would be no apparent change.

Since flight beyond the speed of light was impossible – although for slightly differing reasons – by both the current theories of relativity, neither theory had offered any clue as to what would happen on board a translight ship. They would not allow that any such ship could even exist. The Haertel transformation, on which, in effect, the DFC–3 flew, was non-relativistic: it showed that the apparent elapsed time of a translight journey should be identical in ship-time, and in the time of observers at both ends of the trip.

But since ship and pilot were part of the same system, both covered by the same expression in Haertel's equation, it had never occurred to anyone that the pilot and the ship might keep different times. The notion was ridiculous.

One-and-a-sevenhundredone, one-and-a-sevenhundred-two, one-and-a-sevenhundredthree, one-and-a-sevenhund-redfour . . .

The ship was keeping ship-time, which was identical with observer-time. It would arrive at the Alpha Centauri system in ten months. But the pilot was keeping Garrard-time, and it was beginning to look as though he wasn't going to arrive at all.

It was impossible, but there it was. Something – almost certainly an unsuspected physiological side-effect of the overdrive field on human metabolism, an effect which naturally could not have been detected in the preliminary, robot-piloted tests of the overdrive – had speeded up Garrard's subjective apprehension of time, and had done a thorough job of it.

The second-hand began a slow, preliminary quivering as the calendar's innards began to apply power to it. *Seventy-hundred-forty-one, seventy-hundred-forty-two, seventy-hundred-forty-three . . .*

At the count of 7,058 the second-hand began the jump to the next graduation. It took it several apparent minutes to get across the tiny distance, and several more to come completely to rest. Later still, the sound came to him:

Pock.

In a fever of thought, but without any real physical agitation, his mind began to manipulate the figures. Since it took him longer to count an individual number as the number became larger, the interval between the two calendar-ticks probably was closer to 7,200 seconds than to 7,058. Figuring backward brought him quickly to the equivalence he wanted:

One second in ship-time was two hours in Garrard-time.

Had he really been counting for what was, for him, two whole hours? There seemed to be no doubt about it. It looked like a long trip ahead.

Just how long it was going to be struck him with stunning force. Time had been slowed for him by a factor of 7,200. He would get to Alpha Centauri in just 720,000 months.

Which was—

Six thousand years!

II

Garrard sat motionless for a long time after that, the Nessus-shirt of warm sweat swathing him persistently, refusing even to cool. There was, after all, no hurry.

Six thousand years. There would be food and water and air for all that time, or for sixty or six hundred thousand years; the ship would synthesize his needs, as a matter of course, for as long as the fuel lasted, and the fuel bred itself. Even if Garrard ate a meal every three seconds of objective, or ship, time (which, he realized suddenly, he wouldn't be able to do, for it took the ship several seconds of objective-time to prepare and serve up a meal once it was ordered; he'd be lucky if he ate once a day, Garrard-time), there would be no reason to fear any shortage of supplies. That had been one of the earliest of the possibilities for disaster that the Project engineers had ruled out in the design of the DFC–3.

But nobody had thought to provide a mechanism which would indefinitely refurbish Garrard. After six thousand years, there would be nothing left of him but a faint film of dust on the DFC–3's dully-gleaming horizontal surfaces. His corpse might outlast him a while, since the ship itself was sterile – but eventually, he would be consumed by the bacteria which he carried in his own digestive tract. He needed that bacterium to synthesize part of his B-vitamin needs while he lived, but it would consume him without compunction once he had ceased to be as complicated and delicately balanced a thing as a pilot – or as any other kind of life.

Garrad was, in short, to die before DFC–3 had gotten fairly away from Sol; and when, after 12,000 apparent-years, the DFC–3 returned to Earth, not even his mummy would be still aboard.

The chill that went through him at that seemed almost

unrelated to the way he thought he felt about the discovery; it lasted an enormously long time, and in so far as he could characterize it at all, it seemed to be a chill of urgency and excitement – not at all the kind of chill he should be feeling at a virtual death-sentence. Luckily it was not as intolerably violent as the last such emotional convulsion; and when it was over, two clock-ticks later, it left behind a residuum of doubt.

Suppose that this effect of time-stretching was only mental? The rest of his bodily-processes might still be keeping ship-time; Garrard had no immediate reason to believe otherwise. If so, he would be able to move about only on ship-time, too; it would take many apparent-months to complete the simplest task.

But he would live, if that were the case. His mind would arrive at Alpha Centauri six thousand years older, and perhaps madder, than his body, but he would live.

If, on the other hand, his bodily movements were going to be as fast as his mental processes, he would have to be enormously careful. He would have to move slowly and exert as little force as possible. The normal human hand movement, in such a task as lifting a pencil, took the pencil from a state of rest to another state of rest – by imparting to it an acceleration of about two feet per second per second – and, of course, decelerated it by the same amount. If Garrard were to attempt to impart to a two-pound weight, which was keeping ship-time, an acceleration of 14,440 ft./sec.^{2} in his time he'd have to exert a force of 900 pounds on it.

The point was not that it couldn't be done – but that it would take as much effort as pushing a stalled jeep. He'd never be able to lift that pencil with his forearm muscles alone; he'd have to put his back into the task.

And the human body wasn't engineered to maintain stresses of that magnitude indefinitely. Not even the most powerful professional weightlifter is forced to show his prowess throughout every minute of every day.

Pock.

That was the calendar again; another second had gone by.
Or another two hours. It had certainly seemed longer than a
second, but less than two hours, too. Evidently subjective-
time was an intensively recomplicated measure. Even in this
world of micro-time – in which Garrard's mind, at least,
seemed to be operating – he could make the lapses between
calendar-ticks seem a little shorter by becoming actively
interested in some problem or other. That would help,
during the waking hours, but it would help only if the rest of
his body were *not* keeping the same time as his mind. If it
were not, then he would lead an incredibly active, but
perhaps not intolerable mental life during the many centuries of his awake-time, and would be mercifully asleep for
nearly as long.

Both problems – that of how much force he could exert
with his body, and how long he could hope to be asleep in
his mind – emerged simultaneously into the forefront of his
consciousness while he still sat inertly on the hammock,
their terms still much muddled together. After the single tick
of the calendar, the ship – or the part of it that Garrard
could see from here – settled back into complete rigidity.
The sound of the engines, too, did not seem to vary in frequency or amplitude, at least as far as his ears could tell. He
was still not breathing. Nothing moved, nothing changed.

It was the fact that he could still detect no motion of his
diaphragm or his rib-cage that decided him at last. His body
had to be keeping ship-time, otherwise he would have
blacked out from oxygen-starvation long before now. That
assumption explained, too, those two incredibly prolonged,
seemingly sourceless saturnalias of emotion through which
he had suffered: they had been nothing more nor less than
the response of his endocrine glands to the purely intellectual reactions he had experienced earlier. He had discovered
that he was not breathing, had felt a flash of panic and had
tried to sit up. Long after his mind had forgotten those two

impulses, they had inched their way from his brain down his nerves to the glands and muscles involved, and actual, *physical* panic had supervened. When that was over, he actually *was* sitting up, though the flood of adrenalin had prevented his noticing the motion as he had made it. The later chill – less violent, and apparently associated with the discovery that he might die long before the trip was completed – actually had been his body's response to a much earlier mental command – the abstract fever of interest he had felt while computing the time-differential had been responsible for it.

Obviously, he was going to have to be very careful with apparently cold and intellectual impulses of any kind – or he would pay for them later with a prolonged and agonizing glandular reaction. Nevertheless, the discovery gave him considerable satisfaction, and Garrard allowed it free play; it certainly could not hurt him to feel pleased for a few hours, and the glandular pleasure might even prove helpful if it caught him at a moment of mental depression. Six thousand years, after all, provided a considerable number of opportunities for feeling down in the mouth; so it would be best to encourage all pleasure-moments, and let the after-reaction last as long as it might. It would be the instants of panic, of fear, of gloom which he would have to regulate sternly the moment they came into his mind; it would be those which would otherwise plunge him into four, five, six, perhaps even ten Garrard-hours of emotional inferno.

Pock.

There now, that was very good: there had been two Garrard-hours which he had passed with virtually no difficulty of any kind, and without being especially conscious of their passage. If he could really settle down and become used to this kind of scheduling, the trip might not be as bad as he had at first feared. Sleep would take immense bites out of it; and during the waking periods he could put in one hell of a lot of creative thinking. During a single day of ship-time, Garrard could get in more thinking than any philosopher of

Earth could have managed during an entire lifetime. Garrard could, if he disciplined himself sufficiently, devote his mind for a century to running down the consequences of a single thought, down to the last detail, and still have millennia left to go on to the next thought. What panoplies of pure reason could he not have assembled by the time 6,000 years had gone by? With sufficient concentration, he might come up with the solution to the Problem of Evil between breakfast and dinner of a single ship's day, and in a ship's month might put his finger on the First Course!

Pock.

Not that Garrard was sanguine enough to expect that he would remain logical or even sane throughout the trip. The vista was still grim, in much of its detail. But the opportunities, too, were there. He felt a momentary regret that it hadn't been Haertel, rather than himself, who had been given such an opportunity—

Pock.

—for the old man could certainly have made better use of it than Garrard could. The situation demanded someone trained in the highest rigours of mathematics to be put to the best conceivable use. Still and all Garrard began to feel—

Pock.

—that he would give a good account of himself, and it tickled him to realize that (as long as he held on to his essential sanity) he would return—

Pock.

—to Earth after ten Earth months with knowledge centuries advanced beyond anything—

Pock.

—that Haertel knew, or that anyone could know—

Pock.

—who had to work within a normal lifetime. *Pck.* The whole prospect tickled him. *Pck.* Even the clock-tick seemed more cheerful. *Pck.* He felt fairly safe now *Pck* in disregarding his drilled-in command *Pck* against moving *Pck,* since in any *Pck* event he *Pck* had already *Pck* moved *Pck* without *Pck* being *Pck* harmed *Pck* Pck Pck Pck Pck *pckpckpckpckpckpckpck* . . .

He yawned, stretched, and got up. It wouldn't do to be too pleased, after all. There were certainly many problems that still needed coping with, such as how to keep the impulse toward getting a ship-time task performed going, while his higher centres were following the ramifications of some purely philosophical point. And besides . . .

And besides, he had just moved.

More than that; he had just performed a complicated manoeuvre with his body *in normal time!*

Before Garrard looked at the calendar itself, the message it had been ticking away at him had penetrated. While he had been enjoying the protracted, glandular backwash of his earlier feeling of satisfaction, he had failed to notice, at least consciously, that the calendar was accelerating.

Good-bye, vast ethical systems which would dwarf the Greeks. Good-bye, calculi aeons advanced beyond the spinor-calculus of Dirac. Good-bye, cosmologies by Garrard which would allot the Almighty a job as third-assistant-waterboy in an n-dimensional backfield.

Good-bye, also, to a project he had once tried to undertake in college – to describe and count the positions of love, of which, according to under-the-counter myth, there were supposed to be at least forty-eight. Garrard had never been able to carry his tally beyond twenty, and he had just lost what was probably his last opportunity to try again.

The micro-time in which he had been living had worn off, only a few objective-minutes after the ship had gone into overdrive and he had come out of the anaesthetic. The long intellectual agony, with its glandular counterpoint, had come to nothing. Garrard was now keeping ship-time.

Garrard sat back down on the hammock, uncertain whether to be bitter or relieved. Neither emotion satisfied him in the end; he simply felt unsatisfied. Micro-time had been bad enough while it lasted; but now it was gone, and everything seemed normal. How could so transient a thing have killed Brown and Cellini? They were stable men, more stable, by his own private estimation, than Garrard himself. Yet he had come through it. Was there more to it than this?

And if there was – what, conceivably, could it be?

There was no answer. At his elbow, on the control-chassis which he had thrust aside during that first moment of infinitely-protracted panic, the calendar continued to tick. The engine-noise was gone. His breath came and went in natural rhythm. He felt light and strong. The ship was quiet, calm, unchanging.

The calendar ticked, faster and faster. It reached and passed the first hour, ship-time, of flight in overdrive.

Pock.

Garrard looked up in surprise. The familiar noise, this time, had been the hour-hand jumping one unit. The minute-hand was already sweeping past the past half-hour. The second-hand was whirling like a propeller – and while he watched it, it speeded up to complete invisibility—

Pock.

Another hour. The half-hour already passed. *Pock.* Another hour. *Pock.* Another. *Pock. Pock. Pock, Pock, Pock, Pock, pck-pck-pck-pck-pckpckpckpck* . . .

The hands of the calendar swirled towards invisibility as time ran away with Garrard. Yet the ship did not change. It stayed there, rigid, inviolate, invulnerable. When the date-tumblers reached a speed at which Garrard could no longer read them, he discovered that once more he could not move – and that, although his whole body seemed to be aflutter like that of a humming-bird, nothing coherent was coming to him through his senses. The room was dimming, becoming redder; or no, it was . . .

But he never saw the end of the process, never was allowed to look from the pinnacle of macro-time towards which the Haertel overdrive was taking him.

The pseudo-death took him first.

III

That Garrard did not die completely, and within a comparatively short time after the DFC–3 had gone into overdrive, was due to the purest of accidents; but Garrard did not know that. In fact, he knew nothing at all for an indefinite period, sitting rigid and staring, his metabolism slowed down to next to nothing, his mind almost utterly inactive. From time to time, a single wave of low-level metabolic activity passed through him – what an electrician might have termed a 'maintenance turnover' – in response to the urgings of some occult survival-urge; but these were of so basic a nature as to reach his consciousness not at all. This was the pseudo-death.

When the observer actually arrived, however, Garrard woke. He could make very little sense out of what he saw or felt even now; but one fact was clear: the overdrive was off – and with it thé crazy alterations in time-rates – and there was strong light coming through one of the ports. The first leg of the trip was over. It had been these two changes in his environment which had restored him to life.

The thing (or things) which had restored him to consciousness, however, was – it was what? It made no sense. It was a construction, a rather fragile one, which completely surrounded his hammock. No, it wasn't a construction, but evidently something alive – a living being, organized horizontally, that had arranged itself in a circle about him. No, it was a number of beings. Or a combination of all of these things.

How it had gotten into the ship was a mystery, but there it was. Or there they were.

'How do you hear?' the creature said abruptly. Its voice,

or their voices, came at equal volume from every point in the circle, but not from any particular point in it. Garrard could think of no reason why that should be unusual.

'I—' he said. 'Or we – we hear with our ears. Here.'

His answer, with its unintentionally-long chain of open vowel-sounds, rang ridiculously. He wondered why he was speaking such an odd language.

'We-they wooed to pitch you-yours thiswise,' the creature said. With a thump, a book from the DFC–3's ample library fell to the desk beside the hammock. 'We wooed there and there and there for a many. You are the being-Garrard. We-they are the clinesterton beademung, with all of love.'

'With all of love,' Garrard echoed. The beademung's use of the language they both were speaking was odd; but again Garrard could find no logical reason why the beademung's usage should be considered wrong.

'Are – are you-they from Alpha Centauri?' he said hesitantly.

'Yes, we hear the twin radioceles, that show there beyond the gift-orifices. We-they pitched that the being-Garrard wooed with most adoration these twins and had mind to them, soft and loud alike. How do you hear?'

This time the being-Garrard understood the question. 'I hear Earth,' he said. 'But that is very soft, and does not show.'

'Yes,' said the beademung. 'It is a harmony, not a first, as ours. The All-Devouring listens to lovers there, not on the radioceles. Let me-mine pitch you-yours so to have mind of the rodalent beademung and other brothers and lovers, along the channel which is fragrant to the being-Garrard.'

Garrard found that he understood the speech without difficulty. The thought occurred to him that to understand a language on its own terms – without having to put it back into English in one's own mind – is an ability that is won only with difficulty and long practice. Yet, instantly his mind said, 'But it *is* English,' which of course it was. The offer the

clinesterton beademung had just made was enormously
hearted, and he in turn was much minded and of love, to his
own delighting as well as to the beademungen; that almost
went without saying.

There were many matings of ships after that, and the
being-Garrard pitched the harmonies of the beademungen,
leaving his ship with the many gift orifices in harmonic for
the All-Devouring to love, while the beademungen made
show of they-theirs.

He tried, also, to tell how he was out of love with the
overdrive, which wooed only spaces and times, and made
featurelings. The rodalent beademung wooed the overdrive,
but it did not pitch he-them.

Then the being-Garrard knew that all the time was de-
voured, and he must hear Earth again.

'I pitch you-them to fullest love,' he told the beademungen, 'I shall adore the radioceles of Alpha and Proxima
Centauri, "on Earth as it is in Heaven". Now the overdrive
my-other must woo and win me, and make me adore a
featureling much like silence.'

'But you will be pitched again,' the clinesterton beademung said. 'After you have adored Earth. You are much
loved by Time, the All-Devouring. We-they shall wait for
this othering.'

Privately Garrard did not faith as much, but he said, 'Yes,
we-they will make a new wooing of the beademungen at
some other radiant. With all of love.'

On this the beademungen made the pitched adorations,
and in the midst the overdrive cut in. The ship with the
many gift orifices and the being-Garrard him-other saw the
twin radioceles sundered away.

Then, once more, came the pseudo-death.

IV

When the small candle lit in the endless cavern of Garrard's pseudo-dead mind, the DFC–3 was well inside the orbit of Uranus. Since the sun was still very small and distant, it made no spectacular display through the nearby port, and nothing called him from the post-death sleep for nearly two days.

The computers waited patiently for him. They were no longer immune to his control; he could now tool the ship back to Earth himself if he so desired. But the computers were also designed to take into account the fact that he might be truly dead by the time the DFC–3 got back. After giving him a solid week, during which time he did nothing but sleep, they took over again. Radio signals began to go out, tuned to a special channel.

An hour later, a very weak signal came back. It was only a directional signal, and it made no sound inside the DFC–3 – but it was sufficient to put the big ship in motion again.

It was that which woke Garrard. His conscious mind was still glazed over with the icy spume of the pseudo-death; and as far as he could see the interior of the cabin had not changed one whit, except for the book on the deck—

The book. The clinesterton beademung had dropped it there. But what under God was a clinesterton beademung? And what was he, Garrard, crying about? It didn't make sense. He remembered dimly some kind of experience out there by the Centauri twins—

—the twin radioceles—

There was another one of those words. It seemed to have Greek roots, but he knew no Greek – and besides, why would Centaurians speak Greek?

He leaned forward and actuated the switch which would roll the shutter off the front port, actually a telescope with a translucent viewing-screen. It showed a few stars, and a faint nimbus off on one edge which might be the Sun. At about one o'clock on the screen, was a planet about the size of a pea which had tiny projections, like teacup handles, on each side. The DFC–3 hadn't passed Saturn on its way out; at that time it had been on the other side of the Sun from the route the starship had had to follow. But the planet was certainly difficult to mistake.

Garrard was on his way home – and he was still alive and sane. Or was he still sane? These fantasies about Centaurians – which still seemed to have such a profound emotional effect upon him – did not argue very well for the stability of his mind.

But they were fading rapidly. When he discovered, clutching at the handiest fragments of the 'memories', that the plural of *beademung* was *beademungen*, he stopped taking the problem seriously. Obviously a race of Centaurians who spoke Greek wouldn't also be forming weak German plurals. The whole business had obviously been thrown up by his unconscious.

But what *had* he found by the Centaurus stars?

There was no answer to that question but that incomprehensible garble about love, the All-Devouring, and beademungen. Possibly, he had never seen the Centaurus stars at all, but had been lying here, cold as a mackerel, for the entire twenty months.

Or had it been 12,000 years? After the tricks the overdrive had played with time, there was no way to tell what the objective date actually was. Frantically Garrard put the telescope into action. Where was the Earth? After 12,000 years—

The Earth was there. Which, he realized swiftly, proved nothing. The Earth had lasted for many millions of years; 12,000 years was nothing to a planet. The Moon was there, too; both were plainly visible, on the far side of the Sun – but not too far to pick them out clearly, with the telescope at highest power. Garrard could even see a clear sun-highlight

on the Atlantic Ocean, not far east of Greenland; evidently
the computers were bringing the DFC–3 in on the Earth
from about 23° north of the plane of the ecliptic.

The Moon, too, had not changed. He could even see on its
face the huge splash of white, mimicking the sun-highlight
on Earth's ocean, which was the magnesium-hydroxide
landing-beacon, which had been dusted over the Mare
Vaporum in the earliest days of space flight, with a dark spot
on its southern edge which could only be the crater Mon-
ilius.

But that again proved nothing. The Moon never changed.
A film of dust laid down by modern man on its face would
last for millennia – what, after all, existed on the Moon to
blow it away? The Mare Vaporum beacon covered more
than 4,000 square miles; age would not dim it, nor could
man himself undo it – either accidentally, or on purpose – in
anything under a century. When you dust an area that large
on a world without atmosphere, it stays dusted.

He checked the stars against his charts. They hadn't
moved; why should they have, in only 12,000 years? The
pointer-stars in the Dipper still pointed to Polaris. Draco, like
a fantastic bit of tape, wound between the two Bears, and
Cepheus and Cassiopeia, as it always had done. These con-
stellations told him only that it was spring in the northern
hemisphere of Earth.

But spring of what year?

Then, suddenly, it occurred to Garrard that he had a
method of finding the answer. The Moon causes tides in the
Earth, and action and reaction are always equal and opposite. The Moon cannot move things on Earth without
itself being affected – and that effect shows up in the moon's
angular momentum. The Moon's distance from the Earth
increases steadily by 0.6 inches every year. At the end of
12,000 years, it should be 600 feet farther away from the
Earth than it had been when Garrard left it.

Was it possible to measure? Garrard doubted it, but he

got out his ephemeris and his dividers anyhow, and took pictures. While he worked, the Earth grew nearer. By the time he had finished his first calculation – which was indecisive, because it allowed a margin of error greater than the distances he was trying to check – Earth and Moon were close enough in the telescope to permit much more accurate measurements.

Which were, he realized wryly, quite unnecessary. The computer had brought the DFC–3 back, not to an observed sun or planet, but simply to a calculated point. That Earth and Moon would not be near that point when the DFC–3 returned was not an assumption that the computer could make. That the Earth was visible from here was already good and sufficient proof that no more time had elapsed than had been calculated for from the beginning.

This was hardly new to Garrard; it had simply been retired to the back of his mind. Actually he had been doing all this figuring for one reason, and one reason only: because deep in his brain, set to work by himself, there was a mechanism that demanded counting. Long ago, while he was still trying to time the ship's calendar, he had initiated compulsive counting – and it appeared that he had been counting ever since. That had been one of the known dangers of deliberately starting such a mental mechanism; and now it was bearing fruit in these perfectly-useless astronomical exercises.

The insight was healing. He finished the figures roughly, and that unheard moron deep inside his brain stopped counting at last. It had been pawing its abacus for twenty months now, and Garrard imagined that it was as glad to be retired as he was to feel it go.

His radio squawked, and said anxiously, 'DFC–3, DFC–3. Garrard, do you hear me? Are you still alive? Everybody's going wild down here. Garrard, if you hear me, call us!'

It was Haertel's voice. Garrard closed the dividers so convulsively that one of the points nipped into the heel of his hand. 'Haertel, I'm here. DFC–3 to the Project. This is Garrard.' And then, without knowing quite why, he added: 'With all of love.'

Haertel, after all the hoopla was over, was more than interested in the time-effects. 'It certainly enlarges the manifold in which I was working,' he said. 'But I think we can account for it in the transformation. Perhaps even factor it out, which would eliminate it as far as the pilot is concerned. We'll see, anyhow.'

Garrard swirled his highball reflectively. In Haertel's cramped old office, in the Project's administration-shack, he felt both strange and as old, as compressed, constricted. He said, 'I don't think I'd do that, Adolph. I think it saved my life.'

'How?'

'I told you that I seemed to die after a while. Since I got home, I've been reading, and I've discovered that the psychologists take far less stock in the individuality of the human psyche than you and I do. You and I are physical scientists, so we think about the world as being all outside our skins – something which is to be observed, but which doesn't alter the essential *I*. But evidently, that old solipsistic position isn't quite true. Our very personalities, really, depend in large part upon *all* the things in our environment, large and small, that exist outside our skins. If by some means you could cut a human being off from every sense-impression that comes to him from outside, he would cease to exist as a personality within two or three minutes. Probably he would die.'

'Unquote: Harry Stack Sullivan,' Haertel said dryly. 'So?'

'So,' Garrard said, 'think of what a monotonous environment the inside of a spaceship is. In ordinary interplanetary flight, in such an environment, even the most hardened spaceman may go off his rocker now and then. You know the typical spaceman's psychosis as well as I do, I suppose.

The man's personality goes rigid, just like his surroundings. Usually he recovers as soon as he makes port, and makes contact with a more or less normal world again.

'But in the DFC–3, I was cut off from the world around me much more severely. I couldn't look outside the ports – I was in overdrive, and there was nothing to see. I couldn't communicate with home, because I was going faster than light. And then I found I couldn't move, too, for an enormous long while; and that even the instruments that are in constant change for the usual spaceman wouldn't be in motion for me. Even those were fixed.

'After the time-rate began to pick up, I found myself in an even more impossible box. The instruments moved, all right, but then they moved too *fast* for me to read them. The whole situation was now utterly rigid – and, in effect, I died. I froze as solid as the ship around me, and stayed that way as long as the overdrive was on.'

'By that showing,' Haertel said drily, 'the time-effects were hardly your friends.'

'But they were, Adolph. Look. Your engines act on subjective-time; they keep it varying along continuous curves – from far-too-slow to far-too-fast – and, I suppose, back down again. Now, this is a *situation of continuous change*. It wasn't marked enough, in the long run, to keep me out of pseudo-death; but it was sufficient to protect me from being obliterated altogether, which I think is what happened to Brown and Cellini. Those men knew that they could shut down the overdrive if they could just get to it, and they killed themselves trying. But I knew that I just had to sit and take it – and, by my great good luck, your sine-curve time-variation made it possible for me to survive.'

'Ah, ha,' Haertel said. 'A point worth considering – though I doubt that it will make interstellar travel very popular!'

He dropped back into silence, his thin mouth pursed. Garrard took a grateful pull at his drink. At last Haertel said:

'Why are you in trouble over these Centaurians? It seems to me that you have done a good job. It was nothing that you were a hero – any fool can be brave – but I see also that you *thought*, where Brown and Cellini evidently only reacted. Is there some secret about what you found when you reached those two stars?'

Garrard said, 'Yes, there is. But I've already told you what it is. When I came out of the pseudo-death, I was just a sort of plastic palimpsest upon which anybody could have made a mark. My own environment, my ordinary Earth environment, was a hell of a long way off. My present surroundings were nearly as rigid as they had ever been. When I met the Centaurians – if I did, and I'm not at all sure of that – *they* became the most important thing in the world, and my personality changed to accommodate and understand them. That was a change about which I couldn't do a thing.

'Possibly I did understand them. But the man who understood them wasn't the same man you're talking to now, Adolph. Now that I'm back on Earth, I don't understand that man. He even spoke English in a way that's gibberish to me. If I can't understand myself during that period – and I can't; I don't even believe that that man was the Garrard I know – what hope have I of telling you or the Project about the Centaurians? They found me in a controlled environment, and they altered me by entering it. Now that they're gone, nothing comes through; I don't even understand why I think they spoke English!'

'Did they have a name for themselves?'

'Sure,' Garrard said. 'They were the beademungen.'

'What did they look like?'

'I never saw them.'

Haertel leaned forward. 'Then—'

'I heard them. I think.' Garrard shrugged, and tasted his Scotch again. He was home, and on the whole he was pleased.

But in his malleable mind he heard someone say, '*On*

Earth, as it is in Heaven,' and then, in another voice, which might also have been his own (why had he thought 'him-other'?), '*It is later than you think.*'

'Adolph,' he said, 'is this all there is to it? Or are we going to go on with it from here? How long will it take to make a better starship, a DFC–4?'

'Many years,' Haertel said, smiling kindly. 'Don't be anxious, Garrard. You've come back, which is more than the others managed to do, and nobody will ask you to go out again. I really think that it's hardly likely that we'll get another ship built during your lifetime; and even if we do, we'll be slow to launch it. We really have very little information about what kind of a playground you found out there.

'I'll go,' Garrard said. 'I'm not afraid to go back – I'd like to go. Now that I know how the DFC–3 behaves, I could take it out again, bring you back proper maps, tapes, photos.'

'Do you really think,' Haertel said, his face suddenly serious, 'that we could let the DFC–3 go out again? Garrard, we're going to take that ship apart practically molecule by molecule; that's preliminary to the building of any DFC–4. And no more can we let you go. I don't mean to be cruel, but has it occurred to you that this desire to go back may be the result of some kind of post-hypnotic suggestion? If so, the more badly you want to go back, the more dangerous to us all you may be. We are going to have to examine you just as thoroughly as we do the ship. If these beademungen wanted you to come back, they must have had a reason – and we have to know that reason.'

Garrard nodded, but he knew that Haertel could see the slight movement of his eyebrows and the wrinkles forming in his forehead, the contractions of the small muscles which stop the flow of tears only to make grief patent on the rest of the face.

'In short,' he said, '*don't move.*'

Haertel looked politely puzzled. Garrard, however, could say nothing more. He had returned to humanity's common time, and would never leave it again.

Not even, for all his dimly-remembered promise, with all there was left in him of love.

A Work of Art

Instantly, he remembered dying. He remembered it, however, as if at two removes – as though he were remembering a memory, rather than an actual event; as though he himself had not really been there when he died.

Yet the memory was all from his own point of view, not that of some detached and disembodied observer which might have been his soul. He had been most conscious of the rasping, unevenly-drawn movements of the air in his chest. Blurring rapidly, the doctor's face had bent over him, loomed, come closer, and then had vanished as the doctor's head passed below his cone of vision, turned sideways to listen to his lungs.

It had become rapidly darker, and then, only then, had he realized that these were to be his last minutes. He had tried dutifully to say Pauline's name, but his memory contained no record of the sound – only of the rattling breath, and of the film of sootiness thickening in the air, blotting out everything for an instant.

Only an instant, and then the memory was over. The room was bright again, and the ceiling, he noticed with wonder, had turned a soft green. The doctor's head lifted again and looked down at him.

It was a different doctor. This one was a far younger man, with an ascetic face and gleaming, almost fey eyes. There was no doubt about it. One of the last conscious thoughts he had had was that of gratitude that the attending physician, there at the end, had not been the one who secretly hated him for his one-time associations with the Nazi hierarchy. The attending doctor, instead, had worn an expression amusingly proper for that of a Swiss expert called to the deathbed of an eminent man: a mixture of worry at the prospect of losing so eminent a patient, and complacency at the thought that at the old man's age, nobody could blame this doctor if he died. At eighty-five, pneumonia is a serious matter, with or without penicillin.

'You're all right now,' the new doctor said, freeing his patient's head of a whole series of little silver rods which had been clinging to it by a sort of network cap. 'Rest a minute and try to be calm. Do you know your name?'

He drew a cautious breath. There seemed to be nothing at all the matter with his lungs now; indeed, he felt positively healthy. 'Certainly,' he said, a little nettled. 'Do you know yours?'

The doctor smiled crookedly. 'You're in character, it appears,' he said. 'My name is Barkun Kris; I am a mind sculptor. Yours?'

'Richard Strauss.'

'Very good,' Dr Kris said, and turned away. Strauss, however, had already been diverted by a new singularity. *Strauss* is a word as well as a name in German; it has many meanings – an ostrich, a bouquet – von Wolzogen had had a high old time working all the possible puns into the libretto of *Feuersnot*. And it happened to be the first German word to be spoken either by himself or by Dr Kris since that twice-removed moment of death. The language was not French or Italian, either. It was most like English, but not the English Strauss knew; nevertheless, he was having no trouble speaking it and even thinking it.

Well, he thought, *I'll be able to conduct 'The Love of Danae' after all. It isn't every composer who can première his own opera posthumously.* Still, there was something queer about all this – the queerest part of all being that conviction, which would not go away, that he had actually been dead for just a short time. Of course medicine was making great strides, but—

'Explain all this,' he said, lifting himself to one elbow. The bed was different, too, and not nearly as comfortable as the one in which he had died. As for the room, it looked more like a dynamo shed than a sickroom. Had modern medicine taken to reviving its corpses on the floor of the Siemans-Schukert plant?

'In a moment,' Dr Kris said. He finished rolling some machine back into what Strauss impatiently supposed to be its place, and crossed to the pallet. 'Now. There are many things you'll have to take for granted without attempting to understand them, Dr Strauss. Not everything in the world today is explicable in terms of your assumptions. Please bear that in mind.'

'Very well. Proceed.'

'The date,' Dr Kris said, 'is 2161 by your calendar – or, in other words, it is now two hundred and twelve years after your death. Naturally, you'll realize that by this time nothing remains of your body but the bones. The body you have now was volunteered for your use. Before you look into a mirror to see what it's like, remember that its physical difference from the one you were used to is all in your favour. It's in perfect health, not unpleasant for other people to look at, and its physiological age is about fifty.'

A miracle? No, not in this new age, surely. It was simply a work of science. But what a science! This was Nietzsche's eternal recurrence and the immortality of the superman combined into one.

'And where is this?' the composer said.

'In Port York, part of the State of Manhattan, in the United States. You will find the country less changed in some respects than I imagine you anticipate. Other changes, of course, will seem radical to you; but it's hard for me to predict which ones will strike you that way. A certain resilience on your part will bear cultivating.'

'I understand,' Strauss said, sitting up. 'One question,

please; is it still possible for a composer to make a living in this country?'

'Indeed it is,' Dr Kris said, smiling. 'As we expect you to do. It is one of the purposes for which we've – brought you back.'

'I gather, then,' Strauss said somewhat drily, 'that there is still a demand for my music. The critics in the old days—'

'That's not quite how it is,' Dr Kris said. 'I understand some of your work is still played, but frankly I know very little about your current status. My interest is rather—'

A door opened somewhere, and another man came in. He was older and more ponderous than Kriss and had a certain air of academicism; but he too was wearing the oddly-tailored surgeon's gown, and looked upon Kris's patient with the glowing eyes of an artist. 'A success, Kris?' he said. 'Congratulations.'

'They're not in order yet,' Dr Kris said. 'The final proof is what counts. Dr Strauss, if you feel strong enough, Dr Seirds and I would like to ask you some questions. We'd like to make sure your memory is clear.'

'Certainly. Go ahead.'

'According to our records,' Kris said, 'you once knew a man whose initials were RKL; this was while you were conducting at the Vienna *Staatsoper*.' He made the double 'a' at least twice too long, as though German were a dead language he was striving to pronounce in some 'classical' accent. 'What was his name, and who was he?'

'That would be Kurt List – his first name was Richard, but he didn't use it. He was assistant stage manager.'

The two doctors looked at each other. 'Why did you offer to write a new overture to "*The Woman Without a Shadow*", and give the manuscript to the City of Vienna?'

'So I wouldn't have to pay the garbage removal tax on the Maria Theresa villa they had given me.'

'In the back yard of your house at Garmisch-Partenkirchen there was a tombstone. What was written on it?'

Strauss frowned. That was a question he would be happy to be unable to answer. If one is to play childish jokes upon oneself, it's best not to carve them in stone, and put the carving where you can't help seeing it every time you go out to tinker with the Mercedes. 'It says,' he replied wearily, ' "*Sacred to the memory of Guntram, Minnesinger, slain in a horrible way by his father's own symphony orchestra.*" '

'When was "*Guntram*" premièred?'

'In – let me see – 1894, I believe.'

'Where?'

'In Weimar.'

'Who was the leading lady?'

'Pauline de Ahna.'

'What happened to her afterwards?'

'I married her. Is she—' Strauss began anxiously.

'No,' Dr Kris said. 'I'm sorry, but we lack the data to reconstruct more or less ordinary people.'

The composer sighed. He did not know whether to be worried or not. He had loved Pauline, to be sure; on the other hand, it would be pleasant to be able to live the new life without being forced to take off one's shoes every time one entered the house, so as not to scratch the polished hardwood floors. And also pleasant, perhaps, to have two o'clock in the afternoon come by without Pauline's everlasting, '*Richard – jetzt komponiert!*'

'Next question,' he said.

For reasons which Strauss did not understand, but was content to take for granted, he was separated from Drs Kris and Seirds as soon as both were satisfied that the composer's memory was reliable and his health stable. His estate, he was given to understand, had long since been broken up – a sorry end for what had been one of the principle fortunes of Europe – but he was given sufficient money to set up lodgings and resume an active life. He was provided, too, with introductions which proved valuable.

It took no longer than he had expected to adjust to the changes that had taken place in music alone. Music was, he quickly began to suspect, a dying art, which would soon have a status not much above that held by flower-arranging back in what he thought of as his own century. Certainly it couldn't be denied that the trend towards fragmentation, already visible back in his own time, had proceeded almost to completion in 2161.

He paid no more attention to American popular tunes than he had bothered to pay in his previous life. Yet it was evident that their assembly line production methods – all the ballad composers openly used a slide-rule-like device called a Hit Machine – now had their counterparts almost throughout serious music.

The conservatives these days, for instance, were the 12-tone composers – always, in Strauss's opinion, a drily mechanical lot, but never more so than now. Their gods – Berg, Schoenberg, Webern – were looked upon by the concert-going public as great masters, on the abstruse side perhaps, but as worthy of reverence as any of the Three B's.

There was one wing of the conservatives, however, which had gone the 12-tone procedure one better. These men composed what was called 'stochastic music', put together by choosing each individual note by consultation with tables of random numbers. Their bible, their basic text, was a volume called *Operational Aesthetics,* which in turn derived from a discipline called information theory; and not one word of it seemed to touch upon any of the techniques and customs of composition which Strauss knew. The ideal of this group was to produce music which would be 'universal' – that is, wholly devoid of any trace of the composer's individuality, wholly a musical expression of the universal Laws of Chance. The Laws of Chance seemed to have a style of their own, all right; but to Strauss it seemed the style of an idiot child being taught to hammer a flat piano, to keep him from getting into trouble.

By far the largest body of work being produced, however, fell into a category misleadingly called 'science-music'. The term reflected nothing but the titles of the works, which dealt with space flight, time travel, and other subjects of a romantic or an unlikely nature. There was nothing in the least scientific about the music, which consisted of a mélange of clichés and imitations of natural sounds, in which Strauss was horrified to see his own time-distorted and diluted image.

The most popular form of science-music was a nine-minute composition called a concerto, though it bore no resemblance at all to the classical concerto form; it was instead a sort of free rhapsody after Rachmaninoff – long after. A typical one – 'Song of Deep Space' it was called, by somebody named H. Valerion Krafft – began with a loud assault on the tam-tam, after which all the strings rushed up the scale in unison, followed at a respectable distance by the harp and one clarinet in parallel 6/4's. At the top of the scale cymbals were bashed together, *forte possible*, and the whole orchestra launched itself into a major-minor, wailing sort of melody; the whole orchestra, that is, except for the French horns, which were plodding back down the scale again in what was evidently supposed to be a counter melody. The second phrase of the theme was picked up by a solo trumpet with a suggestion of tremolo; the orchestra died back to its roots to await the next cloud burst, and at this point – as any four-year-old could have predicted – the piano entered with the second theme.

Behind the orchestra stood a group of thirty women, ready to come in with a wordless chorus intended to suggest the eeriness of Deep Space – but at this point, too, Strauss had already learned to get up and leave. After a few such experiences he could also count upon meeting in the lobby Sindi Noniss, the agent to which Dr Kris had introduced him, and who was handling the reborn composer's output –

what there was of it thus far. Sindi had come to expect these walkouts on the part of his client, and patiently awaited them, standing beneath a bust of Gian-Carlo Menotti; but he liked them less and less, and lately had been greeting them by turning alternately red and white like a toti-potent barber-pole.

'You shouldn't have done it,' he burst out after the Krafft incident. 'You can't just walk out on a new Krafft composition. The man's the president of the Interplanetary Society for Contemporary Music. How am I ever going to persuade them that you're a contemporary if you keep snubbing them?'

'What does it matter?' Strauss said. 'They don't know me by sight.'

'You're wrong; they know you very well, and they're watching every move you make. You're the first major composer the mind sculptors ever tackled, and the ISCM would be glad to turn you back with a rejection slip.'

'Why?'

'Oh,' said Sindi, 'there are lots of reasons. The sculptors are snobs; so are the ISCM boys. Each of them wants to prove to the other that their own art is the king of them all. And then there's the competition; it would be easier to flunk you than to let you into the market. I really think you'd better go back in. I could make some excuse—'

'No,' Strauss said shortly. 'I have work to do.'

'But that's just the point, Richard. How are we going to get an opera produced without the ISCM? It isn't as though you wrote theremin solos, or something that didn't cost so—'

'I have work to do,' he said, and left.

And he did: work which absorbed him as had no other project during the last thirty years of his former life. He had scarcely touched pen to music paper – both had been astonishingly hard to find – when he had realized that nothing in his long career had provided him with touchstones by which to judge what music he should write *now*.

The old tricks came swarming back by the thousands, to be sure: the sudden, unexpected key-changes at the crest of a melody; the interval stretching; the piling of divided strings, playing in the high harmonics, upon the already tottering top of a climax; the scurry and bustle as phrases were passed like lightning from one choir of the orchestra to another; the flashing runs in the brass, the chuckling in the clarinets, the snarling mixtures of colours to emphasize dramatic tension – all of them.

But none of them satisfied him now. He had been content with them for most of a lifetime, and had made them do an astonishing amount of work. But now it was time to strike out afresh. Some of the tricks, indeed, actively repelled him: where had he gotten the notion, clung to for decades, that violins screaming out in unison somewhere in the stratosphere was a sound interesting enough to be worth repeating inside a single composition, let alone in all of them?

And nobody, he reflected contentedly, ever approached such a new beginning better equipped. In addition to the past lying available in his memory, he had always had a technical armamentarium second to none; even the hostile critics had granted him that. Now that he was, in a sense, composing his first opera – his first after fifteen of them! – he had every opportunity to make it a masterpiece.

And every such intention.

There were, of course, many minor distractions. One of them was that search for old-fashioned score paper, and a pen and ink with which to write on it. Very few of the modern composers, it developed, wrote their music at all. A large bloc of them used tape, patching together snippets of tone and sound snippets from other tapes, superimposing one tape on another, and varying the results by twirling an elaborate array of knobs this way or that. Almost all the composers of 3–V scores, on the other hand, wrote on the sound-track itself, rapidly scribbling jagged wiggly lines which, when passed through a photocell-audio circuit, produced a noise reasonably like an orchestra playing music, overtones and all.

The last-ditch conservatives who still wrote notes on paper, did so with the aid of a musical typewriter. The device, Strauss had to admit, seemed perfected at last; it had manuals and stops like an organ, but it was not much more than twice as large as a standard letter-writing typewriter, and produced a neat page. But he was satisfied with his own spidery, highly-legible manuscript and refused to abandon it, badly though the one pen-nib he had been able to buy coarsened it. It helped to tie him to his past.

Joining the ISCM had also caused him some bad moment, even after Sindi had worked him around the political roadblocks. The Society man who examined his qualifications as a member had run through the questions with no more interest than might have been shown by a veterinarian examining his four thousandth sick calf.

'Had anything published?'

'Yes, nine tone-poems, about three hundred songs, an—'

'Not when you were alive,' the examiner said, somewhat disquietingly. 'I mean since the sculptors turned you out again.'

'Since the sculptors – ah, I understand. Yes, a string quartet, two song cycles, a—'

'Good. Alfie, write down, "songs". Play an instrument?'

'Piano.'

'Hm.' The examiner studied his fingernails. 'Oh, well. Do you read music? Or do you use a scriber, or tape-clips? Or a Machine?'

'I read.'

'Here.' The examiner sat Strauss down in front of a viewing lectern, over the lit surface of which an endless belt of translucent paper was travelling. On the paper was an immensely magnified sound-track. 'Whistle me the tune of that, and name the instruments it sounds like.'

'I don't read that *Musiksticheln*,' Strauss said frostily, 'or write it, either. I use standard notation, on music paper.'

'Alfie, write down, "Reads notes only." ' He laid a sheet of greyly-printed music on the lectern above the ground glass. 'Whistle me that.'

'That' proved to be a popular tune called 'Vangs, Snifters and Store-Credit Snooky' which had been written on a Hit Machine in 2159 by a guitar-faking politician who sang it at campaign rallies. (In some respects, Strauss reflected, the United States had indeed not changed very much.) It had become so popular that anybody could have whistled it from the title alone, whether he could read the music or not. Strauss whistled it, and to prove his bona fides added, 'It's in the key of B flat.'

The examiner went over to the green-painted upright piano and hit one greasy black key. The instrument was horribly out of tune – the note was much nearer to the standard 440/cps. A than it was to B flat – but the examiner said, 'So it is. Alfie, write down "Also reads flats." All right, son, you're a member. Nice to have you with us; not many people can read that old-style notation any more. A lot of them think they're too good for it.'

'Thank you,' Strauss said.

'My feeling is, if it was good enough for the old masters, it's good enough for us. We don't have people like them with us these days, it seems to me. Except for Dr Krafft, of course, They were *great* back in the old days – men like Shilkret, Steiner, Tiomkin, and Pearl ... and Wilder and Jannsen. Real goffin.'

'*Doch gewiss*,' Strauss said politely.

But the work went forward. He was taking a little income now, from small works. People seemed to feel a special interest in a composer who had come out of the mind sculptors' laboratories; and in addition, the material itself, Strauss was quite certain, had merits of its own to help sell it.

It was the opera which counted, however. That grew and grew under his pen, as fresh and new as his new life, as founded in knowledge and ripeness as his long full memory. Finding a libretto had been troublesome at first. While it was possible that something existed that might have served among the current scripts for 3–V – though he doubted it – he found himself unable to tell the good from the bad through the fog cast over both by incomprehensibly technical production directions. Eventually, and for only the third time in his whole career, he had fallen back upon a play written in a language other than his own, and – for the first time – decided to set it in that language.

The play was Christopher Fry's '*Venus Observed*', in all ways a perfect Strauss opera libretto, as he came gradually to realize. Though nominally a comedy, with a complex farcical plot, it was a verse play with considerable depth to it, and a number of characters who cried out to be brought by music into three dimensions, plus a strong undercurrent of autumnal tragedy, of leaf-fall and apple-fall – precisely the kind of contradictory dramatic mixture which von Hofmannsthal had supplied him in '*The Knight of the Rose*', in '*Ariadne at Naxos*', and in '*Arabella*'.

Alas for von Hofmannsthal, but here was another long-dead playwright who seemed nearly as gifted; and the musical opportunities were immense. There was, for instance, the fire which ended act two; what a gift for a composer to whom orchestration and counterpoint were as important as air and water! Or take the moment where Perpetua shoots the apple from the Duke's hand; in that one moment a single passing reference could add Rossini's marmoreal '*William Tell*' to the musical texture as nothing but an ironic footnote! And the Duke's great curtain speech, beginning:

> *Shall I be sorry for myself? In Mortality's*
> *name*
> *I'll be sorry for myself. Branches and boughs,*
> *Brown hills, the valleys faint with*
> *brume,*
> *Aburnish on the lake; . . .*

There was a speech for a great tragic comedian, in the spirit of Falstaff; the final union of laughter and tears, punctuated by the sleepy comments of Reedbeck, to whose sonorous snore (Trombones, no less than five of them, *con sordini*?) the opera would gently end . . .

What could be better? And yet he had come upon the play only by the unlikeliest series of accidents. At first he had planned to do a straight knockabout farce, in the idiom of '*The Silent Woman*', just to warm himself up. Remembering that Zweig had adapted that libretto for him, in the old days, from a play by Ben Jonson, Strauss had begun to search out English plays of the period just after Jonson's, and had promptly run aground on an awful specimen in heroic couplets called '*Venice Preserv'd*', by one William Atwe. The Fry play had directly followed the Atwe in the card catalogue, and he had looked at it out of curiosity; why should a twentieth-century playwright be punning on a title from the eighteenth?

After two pages of the Fry play, the minor puzzle of the pun disappeared entirely from his concern. His luck was running again; he had an opera.

Sindi worked miracles in arranging for the performance. The date of the première, as set even before the score was finished, reminding Strauss pleasantly of those heady days when Fuerstner had been snatching the conclusion of '*Elektra*' off his work-table a page at a time, before the ink was even dry, to rush it to the engraver before publication deadline. The situation now, however, was even more complicated, for some of the score had to be scribed, some of it taped, some of it engraved in the old way, to meet the new techniques of performance; there were moments when Sindi seemed to be turning quite grey.

But '*Venus Observed*' was, as usual, forthcoming complete from Strauss's pen in plenty of time. Writing the music in first draft had been hellishly hard work, much more like being reborn than had been that confused awakening in Barkun Kris's laboratory, with its overtones of being dead instead; but Strauss found that he still retained all of his old ability to score from the draft almost effortlessly, as undisturbed by Sindi's half-audible worrying in the room with him as he was by the terrifying supersonic bangs of the rockets that bulleted invisibly over the city.

When he was finished, he had two days still to spare before the beginning of rehearsals. With those, furthermore, he would have nothing to do. The techniques of performance in this age were so completely bound up with the electronic arts as to reduce his own experience – he, the master Kapellmeister of them all – to the hopelessly primitive.

He did not mind. The music, as written, would speak for itself. In the meantime he found it grateful to forget the month-long preoccupation with the stage for a while. He went back to the library and browsed lazily through old poems, vaguely seeking texts for a song or two. He knew better than to bother with recent poets; they could not speak to him, and he knew it. The Americans of his own age, he thought, might give him a clue to understanding this America of 2161; and if some such poem gave birth to a song, so much the better.

The search was relaxing and he gave himself up to enjoying it. Finally he struck a tape that he liked: a tape read in a cracked old voice that twanged of Idaho as that voice had

twanged in 1910, in Strauss's own ancient youth. The poet's
name was Pound; he said, on the tape:[1]

> '. . . *the souls of all men great*
> *At times pass through us,*
> *And we are melted into them,*
>   *and are not*
> *Save reflections of their souls.*
> *Thus I am Dante for a space*
>   *and am*
>   *One Francois Villon ballad-*
>   *lord and thief*
> *Or am such holy ones I may*
>   *not write,*
> *Lest Blasphemy be writ against*
>   *my name;*
> *This for an instant and the*
>   *flame is gone.*
> ' *'Tis as in midmost us there*
>   *glows a sphere*
> *Translucent, molten gold, that*
>   *is the 'I'*
> *And into this some form projects*
>   *itself:*
> *Christus, or John, or eke the*
>   *Florentine;*
> *And as the clear space is not*
>   *if a form's*
> *Imposed thereon,*
> *So cease we from all being*
>   *for the time,*
> *And these, the Masters of the*
> *Soul, live on.*'

[1] The full text of this poem may be found in 'Personae' by Ezra
Pound (published by Faber and Faber).

He smiled. That lesson had been written again and again, from Plato onward. Yet the poem was a history of his own case, a sort of theory for the metempsychosis he had undergone, and in its formal way it was moving. It would be fitting to make a little hymn of it, in honour of his own rebirth, and of the poet's insight.

A series of solemn, breathless chords framed themselves in his inner ear, against which the words might be intoned in a high, gently blending hush at the beginning ... and then a dramatic passage in D which the great names of Dante and Villon would enter ringing like challenges to Time ... He wrote for a while in his notebook before he returned the spool to its shell.

These, he thought, are good auspices.

And so, the night of the première arrived, the audience pouring into the hall, the 3-V cameras riding on no visible supports through the air, and Sindi calculating his share of his client's earnings by a complicated game he played on his fingers, the basic law of which seemed to be that one plus one equals ten. The hall filled to the roof with people from every class, as though what was to come would be a circus rather than an opera.

There were, surprisingly, nearly fifty of the aloof and aristocratic mind sculptors, clad in formal clothes which were exaggerated black versions of their surgeon's gowns. They had bought a bloc of seats near the front of the auditorium, where the gigantic 3-V figures which would shortly fill the 'stage' before them (the real singers would perform on a small stage in the basement) could not seem monstrously out of proportion; but Strauss supposed that they had taken this into account and dismissed it.

There was a tide of whispering in the audience as the sculptors began to trickle in, and with it an undercurrent of excitement the meaning of which was unknown to Strauss. He did not attempt to fathom it, however; he was coping with his own mounting tide of opening-night tension, which despite all the years he had never quite been able to shake.

The sourceless, gentle light in the auditorium dimmed, and Strauss mounted the podium. There was a score before him, but he doubted that he would need it. Directly before him, poking up from among the musicians, were the inevitable 3–V snouts, waiting to carry his image to the singers in the basement.

The audience was quiet now. This was the moment. His baton swept up and then decisively down, and the prelude came surging up out of the pit.

For a little while he was deeply immersed in the always tricky business of keeping the enormous orchestra together and sensitive to the flexing of the musical web beneath his hand. As his control firmed and became secure, however, the task became slightly less demanding, and he was able to pay more attention to what the whole sounded like.

There was something decidedly wrong with it. Of course there were the occasional surprises as some bit of orchestral colour emerged with a different *klang* than he had expected; that happened to every composer, even after a lifetime of experience. And there were moments when the singers, entering upon a phrase more difficult to handle than he had calculated, sounded like someone about to fall off a tightrope (although none of them actually fluffed once; they were as fine a troup of voices as he had ever had to work with).

But these were details. It was the overall impression that was wrong. He was losing not only the excitement of the première – after all, that couldn't last at the same pitch all evening – but also his very interest in what was coming from the stage and the pit. He was gradually tiring; his baton arm becoming heavier; as the second act mounted to what should have been an impassioned outpouring of shining tone, he was so bored as to wish he could go back to his desk to work on that song.

Then the act was over; only one more to go. He scarcely heard the applause. The twenty minutes' rest in his dressing-room was just barely enough to give him the necessary strength.

And suddenly, in the middle of the last act, he understood.

There was nothing new about the music. It was the old Strauss all over again – but weaker, more dilute than ever. Compared with the output of composers like Krafft, it doubtless sounded like a masterpiece to this audience. But he knew.

The resolutions, the determination to abandon the old clichés and mannerisms, the decision to say something new – they had all come to nothing against the force of habit. Being brought to life again meant bringing to life as well as those deeply-graven reflexes of his style. He had only to pick up his pen and they overpowered him with easy automatism, no more under his control than the jerk of a finger away from a flame.

His eyes filled; his body was young, but he was an old man, an old man. Another thirty-five years of this? Never. He had said all this before, centuries before. Nearly a half century condemned to saying it all over again, in a weaker and still weaker voice, aware that even this debased century would come to recognize in him only the burnt husk of greatness – no; never.

He was aware, dully, that the opera was over. The audience was screaming its joy. He knew the sound. They had screamed that way when 'Day of Peace' had been premièred, but they had been cheering the man he had been, not the man that 'Day of Peace' showed with cruel clarity he had become. Here the sound was even more meaningless: cheers of ignorance, and that was all.

He turned slowly. With surprise, and with a surprising

sense of relief, he saw that the cheers were not, after all, for him.

They were for Dr Barkun Kris.

Kris was standing in the middle of the bloc of mind sculptors, bowing to the audience. The sculptors nearest him were shaking his hand one after the other. More grasped at it as he made his way to the aisle, and walked forward to the podium. When he mounted the rostrum and took the composer's limp hand, the cheering became delirious.

Kris lifted his arm. The cheering died instantly to an intent hush.

'Thank you,' he said clearly. 'Ladies and gentlemen, before we take leave of Dr Strauss, let us again tell him what a privilege it has been for us to hear this fresh example of his mastery. I am sure no farewell could be more fitting.'

The ovation lasted five minutes, and would have gone on another five if Kris had not cut it off.

'Dr Strauss,' he said, 'in a moment, when I speak a certain formulation to you, you will realize that your name is Jerom Bosch, born in our century and with a life in it all of your own. The superimposed memories which have made you assume the mask, the *persona* of a great composer will be gone. I tell you this so that you may understand why these people here share your applause with me.'

A wave of assenting sound.

'The art of mind sculpture – the creation of artificial personalities for aesthetic enjoyment – may never reach such a pinnacle again. For you should understand that as Jerom Bosch you had no talent for music at all; indeed, we searched a long time to find a man who was utterly unable to carry even the simplest tune. Yet we were unable to impose upon such unpromising material not only the personality, but the genius of a great composer. That genius belongs entirely to you – to the *persona* that thinks of itself as

Richard Strauss. None of the credit goes to the man who
volunteered for the sculpture. That is your triumph, and we
salute you for it.'

Now the ovation could no longer be contained. Strauss,
with a crooked smile, watched Dr Kris bow. This mind
sculpturing was a suitably sophisticated kind of cruelty for
this age; but the impulse that had made Rembrandt and
Leonardo turn cadavers into art-works.

It deserved a suitably sophisticated payment under the *lex
talionis:* an eye for an eye, a tooth for a tooth – and a failure
for a failure.

No, he need not tell Dr Kris that the 'Strauss' he had
created, from nothing but history and old attitudes, was as
empty of genius as a hollow gourd. The joke would always
be on the sculptor, who was incapable of hearing the hollow-
ness of the music now preserved on the 3–V tapes.

But for an instant a surge of revolt poured through his
blood-stream. *I am I,* he thought. *I am Richard Strauss until
I die, and will never be Jerom Bosch, who was utterly unable
to carry even the simplest tune.* His hand, still holding the
baton, came sharply up, though whether to deliver or to
ward off a blow he could not tell.

He let it fall again, and instead, at last, bowed – not to the
audience, but to Dr Kris. He was sorry for nothing, as Kris
turned to him to say the word that would plunge him back
into oblivion, except that he would now have no chance to
set that poem to music.

To Pay the Piper

The man in the white jacket stopped at the door marked 'Re-Education Project – Col. H. H. Mudgett, Commanding Officer' and waited while the scanner looked him over. He had been through that door a thousand times, but the scanner made as elaborate a job of it as if it had never seen him before.

It always did, for there was always in fact a chance that it *had* never seen him before, whatever the fallible human beings to whom it reported might think. It went over him from grey, crew-cut poll to reagent-proof shoes, checking his small wiry body and lean profile against its stored silhouettes, tasting and smelling him as dubiously as if he were an orange held in storage two days too long.

'Name?' it said at last.

'Carson, Samuel, 32–454–0698.'

'Business?'

'Medical director, Re-Ed One.'

While Carson waited, a distant, heavy concussion came rolling down upon him through the mile of solid granite above his head. At the same moment, the letters on the door – and everything else inside his cone of vision – blurred distressingly, and a stab of pure pain went lancing through his head. It was the supersonic component of the explosion, and it was harmless – except that it always both hurt and scared him.

The light on the door-scanner, which had been glowing yellow up to now, flicked back to red again and the machine began the whole routine all over; the sound-bomb had reset it. Carson patiently endured its inspection, gave his name, serial number and mission once more, and this time got the green. He went in, unfolding as he walked the flimsy square of cheap paper he had been carrying all along.

Mudgett looked up from his desk and said at once: 'What now?'

The physician tossed the square of paper down under Mudgett's eyes. 'Summary of the press reaction to Hameline's speech last night,' he said. 'The total effect is going against us, Colonel. Unless we can change Hamelin's mind, this outcry to re-educate civilians ahead of soldiers is going to lose the war for us. The urge to live on the surface again has been mounting for ten years; now it's got a target to focus on. Us.'

Mudgett chewed on a pencil while he read the summary; a blocky, bulky man, as short as Carson and with hair as grey and close-cropped. A year ago, Carson would have told him that nobody in Re-Ed could afford to put stray objects in his mouth even once, let alone as a habit; now Carson just waited. There wasn't a man – or a woman or a child – of America's surviving thirty-five million 'sane' people who didn't have some such tie. Not now, not after twenty-five years of underground life.

'He knows it's impossible, doesn't he?' Mudgett demanded abruptly.

'Of course he doesn't,' Carson said impatiently. 'He doesn't know any more about the real nature of the project than the people do. He thinks the "educating" we do is in some sort of survival technique— That's what the papers think, too, as you can plainly see by the way they loaded that editorial.'

'Um. If we'd taken direct control of the papers in the first place—'

Carson said nothing. Military control of every facet of civilian life was a fact, and Mudgett knew it. He also knew that an appearance of freedom to think is a necessity for the human mind – and that the appearance could not be maintained without a few shreds of the actuality.

'Suppose we do this,' Mudgett said at last. 'Hamelin's position in the State Department makes it impossible for us to muzzle him. But it ought to be possible to explain to him that no unprotected human being can live on the surface, no matter how many Merit Badges he has for woodcraft and first-aid. Maybe we could even take him on a little trip topside; I'll wager he's never seen it.'

'And what if he dies up there?' Carson said stonily. 'We lose three-fifths of every topside party as it is – and Hamelin's an inexperienced—'

'Might be the best thing, mightn't it?'

'*No*,' Carson said. 'It would look like we'd planned it that way. The papers would have the populace boiling by the next morning.'

Mudgett groaned and nibbled another double row of indentations around the barrel of the pencil. 'There must be something,' he said.

'There is.'

'Well?'

'Bring the man here and show him just what we *are* doing. Re-educate *him*, if necessary. Once we told the newspapers that he'd taken the course . . . well, who knows, they just might resent it. Abusing his clearance privileges and so on.'

'We'd be violating our basic policy,' Mudgett said slowly. ' "Give the Earth back to the men who fight for it." Still, the idea has some merits . . .'

'Hamelin is out in the antechamber right now,' Carson said. 'Shall I bring him in?'

The radio-activity never did rise much beyond a mildly hazardous level, and that was only transient, during the second week of the war – the week called the Death of Cities. The small shards of sanity retained by the high commands on both sides dictated avoiding weapons with a built-in back-fire; no cobalt bombs were dropped, no territories

permanently poisoned. Generals still remembered that un-
occupied territory, no matter how devastated, is still un-
conquered territory.

But no such considerations stood in the way of biological
warfare. It was controllable: you never released against the
enemy any disease you didn't yourself know how to control.
There would be some slips, of course, but the margin for
error—

There were some slips. But for the most part, biological
warfare worked fine. The great fevers washed like tides
around and around the globe, one after another. In such
cities as had escaped the bombings, the rumble of truck
convoys carrying the puffed heaped corpses to the mass
graves became the only sound except for sporadic small-
arms fire; and then that too ceased, and the trucks stood
rusting in rows.

Nor were human beings the sole victims. Cattle fevers
were sent out. Wheat rusts, rice moulds, corn blights, hog
choleras, poultry enteritides fountained into the indifferent
air from the hidden laboratories, or were loosed far aloft, in
the jet-stream, by rocketing fleets. Gelatin capsules pullu-
lating with gill-rots fell like hail into the great fishing
grounds of Newfoundland, Oregon, Japan, Sweden,
Portugal. Hundreds of species of animals were drafted as
secondary hosts for human diseases, were injected and released to carry the blessings of the laboratories to their
mates and litters. It was discovered that minute amounts
of the tetrocycline series of antibiotics, which had long
been used as feed supplements to bring farm animals to
full market weight early, could also be used to raise
the most whopping Anopheles and Aëdes mosquitoes anybody ever saw, capable of flying long distances against
the wind and of carrying a peculiarly interesting new
strain of the malarial parasite and the yellow fever
virus . . .

By the time it had ended, everyone who remained alive was a mile underground.

For good.

'I still fail to understand why,' Hamelin said, 'if, as you claim, you have methods of re-educating soldiers for surface life, you can't do so for civilians as well. Or instead.'

The under-secretary, a tall, spare man, bald on top, and with a heavily creased forehead, spoke with the odd neutral accent – untinged by regionalism – of the trained diplomat, despite the fact that there had been no such thing as a foreign service for nearly half a century.

'We're going to try to explain that to you,' Carson said. 'But we thought that, first of all, we'd try to explain once more why we think it would be bad policy – as well as physically out of the question.

'Sure, everybody wants to go topside as soon as it's possible. Even people who are reconciled to these endless caverns and corridors hope for something better for their children – a glimpse of sunlight, a little rain, the fall of a leaf. That's more important now to all of us than the war, which we don't believe in any longer. That doesn't even make any military sense, since we haven't the numerical strength to occupy the enemy's territory any more, and they haven't the strength to occupy ours. We understand all that. But we also know that the enemy is intent on prosecuting the war to the end. Extermination is what they say they want, on their propaganda broadcasts, and your own Department reports that they seem to mean what they say. So we can't give up fighting them; that would be simple suicide. Are you still with me?'

'Yes, but I don't see—'

'Give me a moment more. If we have to continue to fight, we know this much: that the first of the two sides to get men on the surface again – so as to be able to *attack* important targets, not just keep them isolated in seas of plague – will be the side that will bring this war to an end. They know that, too. We have good reason to believe that they have a re-education project, and that it's about as far advanced as ours is.'

'Look at it this way,' Col Mudgett burst in unexpectedly. 'What we have now is a stalemate. A saboteur occasionally locates one of the underground cities and lets the pestilences into it. Sometimes on our side, sometimes on theirs. But that only happens sporadically, and it's just more of this mutual extermination business – to which we're committed, willy-nilly, for as long as they are. If we can get troops on to the surface first, we'll be able to scout out their important installations in short order, and issue them a surrender ultimatum with teeth in it. They'll take it. The only other course is the sort of slow, mutual suicide we've got now.'

Hamelin put the tips of his fingers together. 'You gentlemen lecture me about policy as if I had never heard the word before. I'm familiar with your argument for sending soldiers first. You assume that you're familiar with all of mine for starting with civilians, but you're wrong, because some of them haven't been brought up at all outside the Department. I'm going to tell you some of them, and I think they'll merit your close attention.'

Carson shrugged. 'I'd like nothing better than to be convinced, Mr Secretary. Go ahead.'

'You of all people should know, Dr Carson, how close our underground society is to a psychotic break. To take a single instance, the number of juvenile gangs roaming these corridors of ours has increased 400 per cent since the rumours about the Re-Education Project began to spread. Or another: the number of individual crimes without motive – crimes committed, just to distract the committer from the grinding monotony of the life we all lead – has now passed the total of all other crimes put together.

'And as for actual insanity – of our thirty-five million

people still unhospitalized, there are four million cases *of which we know*, each one of which should be committed right now for early paranoid schizophrenia – except that were we to commit them, our essential industries would suffer a manpower loss more devastating than anything the enemy has inflicted upon us. Every one of those four million persons is a major hazard to his neighbours and to his job, but how can we do without them? And what can we do about the unrecognized, subclinical cases, which probably total twice as many? How long can we continue operating without a collapse under such conditions?'

Carson mopped his brow. 'I didn't suspect that it had gone that far.'

'It has gone that far,' Hamelin said icily, 'and it is accelerating. Your own project has helped to accelerate it. Col. Mudgett here mentioned the opening of isolated cities to the pestilences. Shall I tell you how Louisville fell?'

'A spy again, I suppose,' Mudgett said.

'No, Colonel. Not a spy. A band of – of vigilantes, of mutineers. I'm familiar with your slogan, "The Earth to those who fight for it." Do you know the counter-slogan that's circulating among the people?'

They waited. Hamelin smiled and said: 'Let's die on the surface.'

'They overwhelmed the military detachment there, put the city administration to death, and blew open the shaft to the surface. About a thousand people actually made it to the top. Within twenty-four hours the city was dead – as the ringleaders had been warned would be the outcome. The warning didn't deter them. Nor did it protect the prudent citizens who had no part in the affair.'

Hamelin leaned forward suddenly. 'People won't wait to be told when it's their turn to be re-educated. They'll be tired of waiting, tired to the point of insanity of living at the bottom of a hole. They'll just go.

'And that, gentlemen, will leave the world to the enemy

... or, more likely, the rats. They alone are immune to everything by now.'

There was a long silence. At last Carson said mildly: 'Why aren't *we* immune to everything by now?'

'Eh? Why – the new generations. They've never been exposed.'

'We still have a reservoir of older people who lived through the war: people who had one or several of the new diseases that swept the world, some as many as five, and yet recovered. They still have their immunities; we know; we've tested them. We know from sampling that no new disease has been introduced by either side in over ten years now. Against all the known ones, we have immunization techniques, anti-sera, antibiotics, and so on. I suppose you get your shots every six months like all the rest of us; we should all be very hard to infect now, and such infections as do take should run mild courses.' Carson held the under-secretary's eyes grimly. 'Now, answer me this question: why is it that, despite all these protections, *every single person* in an opened city dies?'

'I don't know,' Hamelin said, staring at each of them in turn. 'By your showing some of them should recover.'

'They should,' Carson said. 'But nobody does. Why? Because the very nature of disease has changed since we all went underground. There are now abroad in the world a number of mutated bacterial strains which can by-pass the immunity mechanisms of the human body altogether. What this means in simple terms is that, should such a germ get into your body, your body wouldn't recognize it as an invader. It would manufacture no antibodies against the germ. Consequently, the germ could multiply without any check, and – you would die. So would we all.'

'I see,' Hamelin said. He seemed to have recovered his composure extraordinarily rapidly. 'I am no scientist, gentlemen, but what you tell me makes our position sound perfectly hopeless. Yet obviously you have some answer.'

Carson nodded. 'We do. But it's important for you to understand the situation, otherwise the answer will mean nothing to you. So: is it perfectly clear to you now, from what we've said so far, that no amount of re-educating a man's brain, be he soldier *or* civilian, will allow him to survive on the surface?'

'Quite clear,' Hamelin said, apparently ungrudgingly. Carson's hopes rose by a fraction of a millimetre. 'But if you don't re-educate his brain, what can you re-educate? His reflexes, perhaps?'

'No,' Carson said. 'His lymph nodes, and his spleen.'

A scornful grin began to appear on Hamelin's thin lips. 'You need better public relations counsel than you've been getting,' he said. 'If what you say is true – as of course I assume it is – then the term "re-educate" is not only inappropriate, it's downright misleading. If you had chosen a less suggestive and more accurate label in the beginning, I wouldn't have been able to cause you half the trouble I have.'

'I agree that we were badly advised there,' Carson said. 'But not entirely for those reasons. Of course the name is misleading; that's both a characteristic and a function of the names of top secret projects. But in this instance, the name 'Re-Education', bad as it now appears, subjected the men who chose it to a fatal temptation. You see, though it is misleading, it is also entirely accurate.'

'Word-games,' Hamelin said.

'Not at all,' Mudgett interposed. 'We were going to spare you the theoretical reasoning behind our project, Mr Secretary, but now you'll just have to sit still for it. The fact is that the body's ability to distinguish between its own cells and those of some foreign tissue – a skin graft, say, or a bacterial invasion of the blood – isn't an inherited ability. It's a learned reaction. Furthermore, if you'll think about it a moment, you'll see that it has to be. Body cells die, too, and have to be disposed of; what would happen if removing those dead cells provoked an antibody reaction, as the destruction of foreign cells does? We'd die of anaphylactic shock while we were still infants.

'For that reason, the body has to learn how to scavenge selectively. In human beings, that lesson isn't learned completely until about a month after birth. During the intervening time, the newborn infant is protected by antibodies that it gets from the colostrum, the "first milk" it gets from the breast during the three or four days immediately after birth. It can't generate its own; it isn't allowed to, so to speak, until it's learned the trick of cleaning up body residues *without* triggering the antibody mechanisms. Any dead cells marked "personal" have to be dealt with some other way.'

'That seems clear enough,' Hamelin said. 'But I don't see its relevance.'

'Well, we're in a position now where that differentiation between the self and everything outside the body doesn't do us any good any more. These mutated bacteria have been "selfed" by the mutation. In other words, some of their protein molecules, probably desoxyribonucleic acid molecules, carry configurations or "recognition-units" identical with those of our body cells, so that the body can't tell one from another.'

'But what has all this to do with re-education?'

'Just this,' Carson said. 'What we do here is to impose upon the cells of the body – all of them – a new set of recognition-units for the guidance of the lymph nodes and the spleen, which are the organs that produce antibodies. The new units are highly complex, and the chances of their being duplicated by bacterial evolution, even under forced draught, are too small to worry about. That's what Re-Education is. In a few moments, if you like, we'll show you just how it's done.'

Hamelin ground out his fifth cigarette in Mudgett's ashtray and placed the tips of his fingers together thoughtfully.

Carson wondered just how much of the concept of recognition-marking the under-secretary had absorbed. It had to be admitted that he was astonishingly quick to take hold of abstract ideas, but the self-marker theory of immunity was – like everything else in immunology – almost impossible to explain to laymen, no matter how intelligent.

'This process,' Hamelin said hesitantly. 'It takes a long time?'

'About six hours per subject, and we can handle only one man at a time. That means that we can count on putting no more than seven thousand troops into the field by the turn of the century. Every one will have to be a highly trained specialist, if we're to bring the war to a quick conclusion.'

'Which means no civilians,' Hamelin said. 'I see. I'm not entirely convinced, but – by all means let's see how it's done.'

Once inside, the under-secretary tried his best to look everywhere at once. The room cut into the rock was roughly two hundred feet high. Most of it was occupied by the bulk of the Re-Education Monitor, a mechanism as tall as a fifteen-storey building, and about a city block square. Guards watched it on all sides, and the face of the machine swarmed with technicians.

'Incredible,' Hamelin murmured. 'That enormous object can process only one man at a time?'

'That's right,' Mudgett said. 'Luckily it doesn't have to treat all the body cells directly. It works through the blood, re-selfing the cells by means of small changes in the serum chemistry.'

'What kind of changes?'

'Well,' Carson said, choosing each word carefully, 'that's more or less a graveyard secret, Mr Secretary. We can tell you this much: the machine uses a vast array of crystalline, complex sugars, which *behave* rather like the blood group-and-type proteins. They're fed into the serum in minute amounts, under feedback control of second-by-second analysis of the blood. The computations involved in deciding upon the amount and the precise nature of each introduced chemical are highly complex. Hence the size of the machine. It is, in its major effect, an artificial kidney.'

'I've seen artificial kidneys in the hospitals,' Hamelin said frowning. 'They're rather compact affairs.'

'Because all they do is remove waste products from the patient's blood, and restore the fluid and electrolyte balance. Those are very minor renal functions in the higher mammals. The organ's main duty is chemical control of immunity. If Burnet and Fenner had known that back in 1949, when the selfing theory was being formulated, we'd have had Re-Education long before now.'

'Most of the machine's size is due to the computation section,' Mudgett emphasized. 'In the body, the brain-stem does those computations, as part of maintaining homeostasis. But we can't reach the brain-stem from outside; it's not under conscious control. Once the body is re-selfed, it will re-train the thalamus where we can't.' Suddenly, two swinging doors at the base of the machine were pushed apart and a mobile operating table came through, guided by two attendants. There was a form on it, covered to the chin with a sheet. The face above this sheet was immobile and almost as white.

Hamelin watched the table go out of the huge cavern with visibly mixed emotions. He said: 'This process – it's painful?'

'No, not exactly,' Carson said. The motive behind the question interested him hugely, but he didn't dare show it. 'But any fooling around with the immunity mechanisms can give rise to symptoms – fever, general malaise, and so on. We try to protect our subjects by giving them a light shock anaesthesia first.'

'Shock?' Hamelin repeated. 'You mean electro shock? I don't see how—'

'Call it stress anaesthesia instead. We give the man a steroid drug that counterfeits the anaesthesia the body itself produces in moments of great stress – on the battlefield, say, or just after a serious injury. It's fast, and free of after-effects. There's no secret about that, by the way; the drug involved is 21-hydroxy-pregnane-3,20-dione sodium succinate, and it dates all the way back to 1955.'

'Oh,' the under-secretary said. The ringing sound of the chemical name had had, as Carson had hoped, a ritually soothing effect.

'Gentlemen,' Hamelin said hesitantly. 'Gentlemen, I have a – a rather unusual request. And, I am afraid, a rather selfish one.' A brief, nervous laugh. 'Selfish in both senses, if you will pardon me the pun. You need feel no hesitation in refusing me, but . . .'

Abruptly he appeared to find it impossible to go on. Carson mentally crossed his fingers and plunged in.

'You would like to undergo the process yourself?' he said.

'Well, yes. Yes, that's exactly it. Does that seem inconsistent? I should know, should I not, what it is that I'm advocating for my following? Know it intimately, from personal experience, not just theory? Of course I realize that it would conflict with your policy, but I assure you I wouldn't turn it to any political advantage – none whatsoever. And perhaps it wouldn't be too great a lapse of policy to process just one civilian among your seven thousand soldiers.'

Subverted, by God! Carson looked at Mudgett with a firmly straight face. It wouldn't do to accept too quickly.

But Hamelin was rushing on, almost chattering now. 'I understand your hesitation. You must feel that I'm trying to gain some advantage, or even to get to the surface ahead of my fellow-men. If it will set your minds at rest, I would be glad to enlist in your advance army. Before five years are up, I could surely learn some technical skill which would make me useful to the expedition. If you would prepare papers to that effect, I'd be happy to sign them.'

'That's hardly necessary,' Mudgett said. 'After you're Re-Educated, we can simply announce the fact, and say that you've agreed to join the advance party when the time comes.'

'Ah,' Hamelin said. 'I see the difficulty. No, that would make my position quite impossible. If there is no other way—'

'Excuse us a moment,' Carson said. Hamelin bowed, and the doctor pulled Mudgett off out of earshot.

'Don't overplay it,' he murmured. 'You're tipping our hand with that talk about a press release, Colonel. He's offering us a bribe – but he's plenty smart enough to see that the price you're suggesting is that of his whole political career; he won't pay that much.'

'What then?' Mudgett whispered hoarsely.

'Get somebody to prepare the kind of informal contract he suggested. Offer to put it under security seal so we won't be able to show it to the press at all. He'll know well enough that such a seal can be broken if our policy ever comes before a presidential review – and that will restrain him from forcing such a review. Let's not demand too much. Once he's been re-educated, he'll have to live the rest of the five years with the knowledge that he *can* live topside any time he wants to try it – and he hasn't had the discipline our men have had. It's my bet that he'll goof off before the five years are up – and good riddance.'

They went back to Hamelin, who was watching the machine and humming in a painfully abstracted manner.

'I've convinced the Colonel,' Carson said, 'that your services in the army might well be very valuable when the time comes, Mr Secretary. If you'll sign up, we'll put the papers under security seal for your own protection, and then I think we can fit you into our treatment programme to-day.'

'I'm grateful to you, Dr Carson,' Hamelin said. 'Very grateful indeed.'

Five minutes after his injection, Hamelin was as peaceful as a flounder and was rolled through the swinging doors. An hour's discussion of the probable outcome, carried on in the privacy of Mudgett's office, bore very little additional fruit, however.

'It's our only course,' Carson said. 'It's what we hoped to gain from his visit, duly modified by circumstances. It all comes down to this: Hamelin's compromised himself, and he knows it.'

'But,' Mudgett said, 'suppose he was right? What about all that talk of his about mass insanity?'

'I'm sure it's true,' Carson said, his voice trembling slightly despite his best efforts at control. 'It's going to be rougher than ever down here for the next five years, Colonel. Our only consolation is that the enemy must have exactly the same problem; and if we can beat them to the surface—'

'*Hsst!*' Mudgett said. Carson had already broken off his sentence. He wondered why the scanner gave a man such a hard time outside that door, and then admitted him without any warning to the people on the other side. Couldn't the damned thing be trained to knock?

The newcomer was a page from the haematology section. 'Here's the preliminary rundown on your "student X", Dr Carson,' he said.

The page saluted Mudgett and went out. Carson began to read. After a moment, he also began to sweat.

'Colonel, look at this. I was wrong after all. Disastrously wrong. I haven't seen a blood-type distribution pattern like Hamelin's since I was a medical student, and even back then it was only a demonstration, not a real live patient. Look at it from the genetic point of view – the migration factors.'

He passed the protocol across the desk. Mudgett was not by background a scientist, but he was an enormously able administrator, of the breed that makes it its business to know the technicalities on which any project ultimately rests. He was not much more than half-way through the tally before his eyebrows were gaining altitude like shock-waves.

'Carson, we can't let that man into the machine! He's—'

'He's already in it, Colonel, you know that. And if we interrupt the process before it runs to term, we'll kill him.'

'Let's kill him, then,' Mudgett said harshly. 'Say he died while being processed. Do the country a favour.'

'That would produce a hell of a stink. Besides, we have no proof.'

Mudgett flourished the protocol excitedly.

'That's not proof to anyone but a haematologist.'

'But Carson, the man's a saboteur!' Mudgett shouted. 'Nobody but an Asiatic could have a typing pattern like this! And he's no melting-pot product, either – he's a classical mixture, very probably a Georgian. And every move he's made since we first heard of him has been aimed directly at us – aimed directly at tricking us into getting him into the machine!'

'I think so too,' Carson said grimly. 'I just hope the enemy hasn't many more agents as brilliant.'

'One's enough,' Mudgett said. 'He's sure to be loaded to the last cc. of his blood with catalyst poisons. Once the machine starts processing his serum, we're done for – it'll take us years to re-programme the computer, if it can be done at all. It's *got* to be stopped!'

'Stopped?' Carson said, astonished. 'But it's already stopped. That's not what worries me. The machine stopped it fifty minutes ago.'

'It can't have! How could it? It has no relevant data!'

'Sure it has.' Carson leaned forward, took the cruelly chewed pencil away from Mudgett, and made a neat check beside one of the entries on the protocol. Mudgett stared at the check item.

'Platelets Rh VI?' he mumbled. 'But what's that got to do

with ... Oh. Oh, I see. That platelet type doesn't exist at all in our population now, does it? Never seen it before myself at least.'

'No,' Carson said, grinning wolfishly. 'It never was common in the West, and the pogrom of 1981 wiped it out. That's something the enemy couldn't know. But the machine knows it. As soon as it gives him the standard anti-IV desensitization shot, his platelets will begin to dissolve – and he'll be rejected for incipient thrombocytopenia.' He laughed. 'For his own protection! But—'

'But he's getting nitrous oxide in the machine, and he'll be held six hours under anaesthesia anyhow – also for his own protection,' Mudgett broke in. He was grinning back at Carson like an idiot. 'When he comes out from under, he'll assume that he's been re-educated, and he'll beat it back to the enemy to report that he's poisoned our machine, so that they can be sure they'll beat us to the surface. And he'll go the fastest way: *overland*.'

'He will,' Carson agreed. 'Of course he'll go overland, and of course he'll die. But where does that leave us? We won't be able to conceal that he was treated here, if there's any sort of inquiry at all. And his death will make everything we do here look like a fraud. Instead of paying our Pied Piper – and great jumping Jehosophat, look at his name! They were rubbing our noses in it all the time! Nevertheless, we didn't pay the piper; we killed him. And "platelets Rh VI" won't be an adequate excuse for the press, or for Hamelin's following.'

'It doesn't worry me,' Mudgett rumbled. 'Who'll know? He won't die in our labs. He'll leave here hale and hearty. He won't die until he makes a break for the surface. After that we can compose a fine obituary for the press. Heroic government official, on the highest policy level – couldn't wait to lead his followers to the surface – died of being too much in a hurry – Re-Ed Project sorrowfully reminds everyone that no technique is foolproof—'

Mudgett paused long enough to light a cigarette, which was a most singular action for a man who never smoked. 'As a matter of fact, Carson,' he said, 'it's a natural.'

Carson considered it. It seemed to hold up. And 'Hamelin' would have a death certificate as complex as he deserved – not officially, of course, but in the minds of everyone who knew the facts. His death, when it came, would be due directly to the thrombocytopenia which had caused the Re-Ed machine to reject him – and thrombocytopenia is a disease of infants. *Unless ye become as little children* ...

That was a fitting reason for rejection from the new kingdom of Earth: anaemia of the newborn.

His pent breath went out of him in a long sigh. He hadn't been aware that he'd been holding it. 'It's true,' he said softly. 'That's the time to pay the piper.'

'When?' Mudgett said.

'When?' Carson said, surprised. 'Why, *before* he takes the children away.'

Nor Iron Bars

I

The *Flyaway II*, which was large enough to carry a hundred passengers, seemed twice as large to Gordon Arpe with only the crew on board – large and silent, with the silence of its orbit a thousand miles above the Earth.

'When are they due?' Dr (now Capt.) Arpe said, for at least the fourth time. His second officer, Freidrich Oestreicher, looked at the chronometer and away again with boredom.

'The first batch will be on board in five minutes,' he said harshly. 'Presumably they've all reached SV–One by now. It only remains to ferry them over.'

Arpe nibbled at a fingernail. Although he had always been the tall, thin, and jumpy type, nail-biting was a new vice to him.

'I still think it's insane to be carrying passengers on a flight like this,' he said.

Oestreicher said nothing. Carrying passengers was no novelty to him. He had been captain of a passenger vessel on the Mars run for ten years, and looked it: a stocky hard-muscled youngster of thirty, whose crew cut was going grey despite the fact that he was five years younger than Arpe. He was second in command of the *Flyaway II* only because he had no knowledge of the new drive. Or, to put it another way, Arpe was captain only because he was the only man who did understand it, having invented it. Either way you put it didn't sweeten it for Oestreicher, that much was evident.

Well, the first officer would be the acting captain most of the time, anyhow. Arpe admitted that he himself had no knowledge of how to run a spaceship. The thought of passengers, furthermore, came close to terrifying him. He hoped to have as little contact with them as possible.

But damn it all, it *was* crazy to be carrying a hundred laymen – half of them women and children, furthermore – on the maiden flight of an untried interstellar drive, solely on the belief of one Dr Gordon Arpe that his brainchild would work. Well, that wasn't the sole reason, of course. The whole Flyaway project, of which Arpe had been head, believed it would work, and so did the government.

And then there was the First Expedition to Centaurus, presumably still in flight after twelve years; they had elected to do it the hard way, on ion drive, despite Garrard's spectacular solo round trip, the Haertel overdrive which had made that possible, being adjudged likely to be damaging to the sanity of a large crew. Arpe's discovery had been a totally unexpected breakthrough, offering the opportunity to rush a new batch of trained specialists to help the First Expedition colonize, arriving only a month or so after the First had landed. And if you are sending help, why not send families, too – the families the First Expedition had left behind?

Which also explained the two crews. One of them consisted of men from the Flyaway project, men who built various parts of the drive, or designed them, or otherwise knew them intimately. The other was made up of men who had served some time – in some cases, as long as two full hitches – in the Space Service under Oestreicher. There was some overlapping, of course. The energy that powered the drive field came from a Nernst-effect generator: a compact ball of fusing hydrogen, held together in mid-combustion chamber by a hard magnetic field, which transformed the heat into electricity to be bled off perpendicular to the magnetic lines of force.[1] The same generator powered the ion

[1] These lines were written before security on the 'magnetic bottle' fusion reactor had been lifted anywhere in the world. They appear here exactly as they did in the original manuscript.—J.B.

rockets of ordinary interplanetary flight, and so could be serviced by ordinary crews. On the other hand, Arpe's new attempt to beat the Lorenz-Fitzgerald equation involved giving the whole ship negative mass, a concept utterly foreign to even the most experienced spaceman. Only a physicist who knew Dirac holes well enough to call them 'Pam' would have thought of the notion at all.

But it would work. Arpe was sure of that. A body with negative mass could come very close to the speed of light before the Fitzgerald contraction caught up with it, and without the wild sine-curve variation in subjective time which the non-Fitzgeraldian Haertel overdrive enforced on the passenger. If the field could be maintained successfully in spite of the contraction, there was no good reason why the velocity of light could not be passed; under such conditions, the ship would not be a material object at all.

And polarity in mass does not behave like polarity in electro-magnetic fields. As gravity shows, where mass is concerned like attracts like, and unlikes repel. The very charging of the field should fling the charged object away from the Earth at a considerable speed.

The unmanned models had not been disappointing. They had vanished instantly, with a noise like a thunderclap. And since every atom in the ship was affected evenly, there ought to be no sensible acceleration, either – which is a primary requirement for an ideal drive. It looked good—

But not for a first test with a hundred passengers!

'Here they come,' said Harold Stauffer, the second officer. Sandy-haired and wiry, he was even younger than Oestreicher, and had the small chin combined with handsome features which is usually called 'a weak face'. He was, Arpe already knew, about as weak as a Diesel locomotive; so much for physiognomy. He was pointing out the viewplate.

Arpe started and followed the pointing finger. At first he saw nothing but the doughnut with the peg in the middle which was Satellite Vehicle 1, as small as a fifty-cent piece at

this distance. Then a tiny silver flame near it disclosed the
first of the ferries, coming towards them.

'We had better get down to the airlock,' Oestreicher said.

'All right,' Arpe responded abstractly. 'Go ahead. I still
have some checking to do.'

'Better delegate it,' Oestreicher said. 'It's traditional for
the captain to meet passengers coming on board. They
expect it. And this batch is probably pretty scared, considering what they've undertaken. I wouldn't depart from
routine with them if I were you, sir.'

'I can run the check,' Stauffer said helpfully. 'If I get into
any trouble on the drive, sir, I can always call your gang
chief. He can be the judge of whether or not to call you.'

Out-generalled, Arpe followed Oestreicher down to the
air-lock.

The first ferry stuck its snub nose into the receiving area;
the nose promptly unscrewed and tipped upward. The first
passenger out was a staggering two-year-old, as bundled up
as though it had been dressed for 'the cold of space', so that
nobody could have told whether it was a boy or a girl. It fell
down promptly, got up again without noticing, and went
charging straight ahead, shouting, 'Bye-bye-see-you, bye-
bye-see-you, bye-bye—' Then it stopped, transfixed, looking
about the huge metal cave with round eyes.

'Judy?' a voice cried from inside the ferry. 'Judy! Judy,
wait for mommy!'

After a moment, the voice's owner emerged: a short, fair
girl, perhaps eighteen. The baby by this time had spotted the
crew member who had the broadest grin, and charged him
shouting, 'Daddy daddy daddy daddy daddy daddy,' like a
machine-gun. The woman followed, blushing.

The crewman was not embarrassed. It was obvious that
he had been called Daddy before by infants on three planets
and five satellites, with what accuracy he might not have
been able to guarantee. He picked up the little girl and
poked her gently.

'Hi-hi, Judy,' he said. 'I see you. Where's Judy? I see her.' Judy crowed and covered her face with her hands; but she was peeking.

'Something's wrong here,' Arpe murmured to Oestreicher. 'How can a man who's been travelling towards Centaurus for twelve years have a two-year-old daughter?'

'Wouldn't raise the question if I were you, sir,' Oestreicher said through motionless lips. 'Passengers are never a uniform lot. Best to get used to it.'

The aphorism was being amply illustrated. Next to leave the ferry was an old woman who might possibly have been the mother of one of the crewmen of the First Centaurus expedition; by ordinary standards she was in no shape to stand a trip through space, and surely she would be no help to anybody when she arrived. She was followed by a striking brunette girl in close-fitting, close-cut leotards, with a figure like a dancer. She might have been anywhere between twenty-one and forty-one years old; she wore no ring, and the hard set of her otherwise lovely face did not suggest that she was anybody's wife. Oddly, she also looked familiar. Arpe nudged Oestreicher and nodded towards her.

'Celia Gospardi,' Oestreicher said out of the corner of his mouth. 'Three-V comedienne. You've seen her, sir, I'm sure.'

And so he had; but he would never have recognized her, for she was not smiling. Her presence here defied any explanation he could imagine.

'Screened or not, there's something irregular about this,' Arpe said in a low voice. 'Obviously there's been a slip in the interviewing. Maybe we can turn some of this lot back.'

Oestreicher shrugged. 'It's your ship, sir,' he said. 'I advise against it, however.'

Arpe scarcely heard him. If some of these passengers were really as unqualified as they looked . . . and there would be no time to send up replacements . . . At random, he started with the little girl's mother.

'Excuse me, ma'am—'

The girl turned with surprise, and then with pleasure. 'Yes, Captain!'

'Uh, it occurs to me that there may have been, uh, an error. The *Flyaway II*'s passengers are strictly restricted to technical colonists and to, uh, legal relatives of the First Centaurus Expedition. Since your Judy looks to be no more than two, and since it's been twelve years since—'

The girl's eyes had already turned ice-blue; she rescued him, after a fashion, from a speech he had suddenly realized he could never have finished. 'Judy,' she said levelly, 'is the granddaughter of Captain Willoughby of the First Expedition. I am his daughter. I am sorry my husband isn't alive to pin your ears back, Captain. Any further questions?'

Arpe left the field without stopping to collect his wounded. He was stopped in mid-retreat by a thirteen-year-old boy wearing astonishingly thick glasses and a thatch of hair that went in all directions in dirty blond cirri.

'Sir,' the boy said, 'I understood that this was to be a new kind of ship. It looks like an SC-Forty-seven freighter to me. Isn't it?'

'Yes,' Arpe said. 'Yes, that's what it is. That is, it's the same hull. I mean, the engines and fittings are new.'

'*Uh*-huh,' the boy said. He turned his back and resumed prowling.

The noise was growing louder as the reception area filled. Arpe was uncomfortably aware that Oestreicher was watching him with something virtually indistinguishable from contempt, but still he could not get away; a small, compact man in a grey suit had hold of his elbow.

'Captain Arpe, I'm Forrest of the President's Commission, to disembark before departure,' he said in a low murmur, so rapidly that one syllable could hardly be told from another. 'We've checked you out and you seem to be in good shape. Just want to remind you that your drive is more important than anything else on board. Get the passengers where they want to go by all means if it's feasible, but if it isn't, *the government wants that drive back*. That means jettisoning the passengers without compunction if necessary. Dig?'

'All right.' That had been pounded into him almost from the beginning of his commission, but suddenly it didn't seem to be as clear-cut a proposition, not now, not after the passengers were actually arriving in the flesh. Filled with a sudden, unticketable emotion, almost like horror, Arpe shook the government man off. Bidding tradition be damned, he got back to the bridge as fast as he could go, leaving Oestreicher to cope with the remaining newcomers. After all, Oestreicher was supposed to know how.

But the rest of the ordeal still loomed ahead of him. The ship could not actually take off until 'tomorrow', after a twelve-hour period during which the passengers would get used to their quarters, and got enough questions answered to prevent their wandering into restricted areas of the ship. And there was still the traditional Captain's Dinner to be faced up to: a necessary ceremony during which the passengers got used to eating in free fall, got rid of their first awkwardness with the tools of space, and got to know each other, with the officers to help them. It was an initial step rather than a final one, as was the Captain's Dinner on the seas.

'Stauffer, how did the check-out go?'

'Mr Stauffer, please, sir,' the second officer said politely. 'All tight, sir. I asked your gang chief to sign the log with me, which he did.'

'Very good. Thank you – uh, Mr Stauffer. Carry on.'

'Yes, sir.'

It looked like a long evening. Maybe Oestreicher would be willing to forego the Captain's Dinner. Somehow, Arpe doubted that he would.

He wasn't willing, of course. He had already arranged for it long ago. Since there was no salon on the converted freighter, the dinner was held in one of the smaller holds, whose cargo had been strapped temporarily in the corridors. The whole inner surface of the hold was taken up by the saddle-shaped tables, to which the guests hitched themselves by belt-hooks; service arrived from way up in the middle of the air.

Arpe's table was populated by the thirteen-year-old boy he had met earlier, a ship's nurse, two technicians from the specialists among the colonist-passengers, a Nernst generator officer, and Celia Gospardi, who sat next to him. Since she had no children of her own with her, she had not been placed at one of the tables allocated to children and parents; besides, she was a celebrity.

Arpe was appalled to discover that she was not the only celebrity on board. At the very next table down was Daryon Hammersmith, the man the newscasts called 'The Conqueror of Titan'. There was no mistaking the huge-shouldered, flamboyant explorer and his heavy voice; he was a natural centre of attention, especially among the women. He was bald, but this simply made him look even more like a Prussian officer of the old school, and as over-poweringly, cruelly masculine as a hunting panther.

For several courses Arpe could think of nothing at all to say. He rather hoped that this blankness of mind would last; maybe the passengers would gather that he was aloof by nature, and ... But the silence at the Captain's table was becoming noticeable, especially against the noise the children were making elsewhere. Next door, Hammersmith appeared to be telling stories.

And what stories! Arpe knew very little about the satellites, but he was somehow quite sure that there were no snow-tigers on Titan who gnawed away the foundations of buildings, nor any three-eyed natives who relished frozen man-meat warmed just until its fluids changed from Ice IV to Ice III. If there were, it was odd that Hammersmith's own book about the Titan expedition had mentioned neither. But the explorer was making Arpe's silence even more conspicuous; he *had* to say something.

'Miss Gospardi – we're honoured to have you with us. You have a husband among the First Expedition, I suppose?'

'Yes, worse luck,' she said, gnawing with even white teeth at a drumstick. 'My fifth.'

'Oh. Well, if at first you don't succeed – isn't that how it goes? You're undertaking quite a journey to be with him again. I'm glad you feel so certain now.'

'I'm certain,' she said calmly. 'It's a long trip, all right. But he made a big mistake when he thought it'd be too long for *me*.'

The thirteen-year-old was watching her like an owl. It looked like a humid night for him.

'Of course, Titan's been tamed down considerably since my time,' Hammersmith was booming jovially. 'I'm told the new dome there is almost cosy, except for the wind. That wind – I still dream about it now and then.'

'I admire your courage,' Arpe said to the 3–V star, beginning to feel faintly courtly. Maybe he had talents he had neglected; he seemed to be doing rather well so far.

'It isn't courage,' the woman said, freeing a piece of bread from the clutches of the Lazy Spider. 'It's desperation. I hate space-flight. I should know, I've had to make that Moon circuit for show dates often enough. But I'm going to get that lousy coward back if it's the last thing I do.'

She took a full third out of the bread-slice in one precise, gargantuan nibble.

'I wouldn't have thought of it if I hadn't lost my sixth husband to Peggy Walton. That skirt-chaser; I must have been out of my mind. But Johnny didn't bother to divorce me before he ran off on this Centaurus safari. That was a mistake. I'm going to haul *him* back by his *scruff*.'

She folded the rest of the bread and snapped it delicately in two. The thirteen-year-old winced and looked away.

'No, I can't say that I miss Titan much,' Hammersmith said, in a meditative tone which nevertheless carried the entire length of the hold. 'I like planets where the sky is clear most of the time. My hobby is micro-astronomy – as a matter of fact I have some small reputation in the field, strictly as an amateur. I understand the stars should be unusually clear and brilliant in the Centaurus area, but of course there's nothing like open space for really serious work.'

'To tell the truth,' Celia went on, although for Arpe's money she had told more than enough truth already, 'I'm scared to death of this bloated coffin of yours. But what the hell, I'm dead anyhow. On Earth, everybody knows I can't stay married two years, no matter how many fan letters I get. Or how many proposals, honourable or natural. It's no good to me any more that three million men say they love me. I know what they mean. Every time I take one of them up on it, he vanishes.'

The folded snippet of bread vanished without a sound.

'Are you really going to be a colonist?' someone asked Hammersmith.

'Not for a while, anyhow,' the explorer said. 'I'm taking my fiancée there—' at least two score feminine faces fell with an almost audible thud – 'to establish our home, but I hope I'll be pushing on ahead with a calibration cruiser. I have a theory that our Captain's drive may involve some navigational difficulties. And I'll be riding my hobby the while; the arrangement suits me nicely.'

Arpe was sure his ears could be seen to be flapping. He was virtually certain that there was no such discipline as microastronomy, and he was perfectly certain that any collimation-cruising (Hammersmith even had the word wrong) the Arpe drive required was going to be done by one Gordon Arpe, except over his dead body.

'*This* man,' Celia Gospardi went on implacably, 'I'm going to hold, if I have to chase him all over the galaxy. 'I'll teach him to run away from *me* without making it legal first.'

Her fork stabbed a heart of lettuce out of the Lazy Spider and turned it in the gout of Russian dressing the Spider had shot into the air after it. 'What does he think he got himself into, anyhow – the Foreign Legion?' she asked nobody in particular. '*Him?* He couldn't find his way out of a supermarket without a map.'

Arpe was gasping like a fish. The girl was smiling warmly at him, from the midst of a cloud of musky perfume against which the ship's ventilators laboured in vain. He had never felt less like the captain of a great ship. In another second he would be squirming. He was already blushing.

'Sir—'

It was Oestreicher, bending at his ear. Arpe almost broke his tether with gratitude. 'Yes, Mr Oestreicher?'

'We're ready to start dogging down; SV-One has asked us to clear the area a little early, in view of the heavy traffic involved. If you could excuse yourself, we're needed on the bridge.'

'Very good. Ladies and gentlemen, please excuse me; I have duties. I hope you'll see the dinner through, and have a good time.'

'Is something wrong?' Celia Gospardi said, looking directly into his eyes. His heart went *boomp!* like a form-stamper.

'Nothing wrong,' Oestreicher said smoothly from behind him. 'There's always work to do in officers' country. Ready, Captain?'

Arpe kicked himself away from the table into the air, avoiding a floating steward only by a few inches. Oestreicher caught up with him in time to prevent his running head-on into the side of a bulkhead.

'We've allowed two hours for the passengers to finish eating and bed down,' Oestreicher reported in the control room. 'Then we'll start building the field. You're sure we don't need any preparations against acceleration?'

Arpe was recovering; now that the questions were technical, he knew where he was. 'No, none at all. The field doesn't mean a thing while it's building. It has to reach a threshold before it takes effect. Once it crosses that point on the curve, it takes effect totally, all at once. Nobody should feel a thing.'

'Good. Then we can hit the hammocks for a few hours. I suggest, sir, that Mr Stauffer take the first watch; I'll take the second; that will leave you on deck when the drive actually fires, if it can be delayed that long. I already have us on a slight retro curve from SV-One.'

'It can be delayed as long as we like. It won't cross the threshold till we close that key.'

'That was my understanding,' Oestreicher said. 'Very good, sir. Then let's stand the usual watches and get under way at the fixed time. By then we'll be at apogee so far as the satellite station is concerned. It would be best to observe normal routine, right up to the moment when the voyage itself becomes unavoidably abnormal.'

This was wisdom, of course. Arpe could do nothing but nod, though he doubted very much that he would manage to get to sleep before his trick came up. The bridge emptied, except for Stauffer and a j.g. from the Nernst gang, and the ship quieted.

In the morning, while the passengers were still asleep, Arpe closed the key.

The *Flyaway II* vanished without a sound.

II

Mommy mommy mommy mommy mommy mommy

I dream I see him   Johnny I love you   he's going down the ladder into the pit and I can't follow and he's gone already and it's time for the next act

Spaceship   I'm flying it and Bobby can see me and all
the people

Some kind of emergency   but then why not the alarms
Got to ring Stauffer

Daddy? Daddy? Bye-bye-see you? Daddy

Where's the bottle   I knew I shouldn't of gotten sucked
into that game

The wind   always the wind

Falling   falling   why can't I stop falling   will I die if I
stop

Two point eight three four. Two point eight three four. I
keep thinking two point eight three four   that's what the
meter says two point eight three four

Somebody stop that wind   I tell you it talks   I tell you I
hear it   words in the wind

Johnny don't go   I'm riding an elephant and he's trying
to go down the ladder after you and it's going to break

No alarms. All well. But can't think. Can't mommy
ladder spaceship think for bye-bye-see-you two windy daddy
bottle seconds straight. What's the bottle trouble game
matter anyhow? Where's that two point eight three four
physicist, what's-his-bye bye name, Daddy, Johnny, Arpe!

will I die if I stop

I love you

the wind

two point

mommy

STOP.

STOP. STOP. Arpe. Arpe. Where are you? Everyone else,
stop thinking. STOP. We're reading each other's minds.
Everyone try to stop before we go nuts. Captain Arpe,
do you hear me? Come to the bridge. Arpe, do you hear
me?

I hear you. I'm on my way. My God.

You there at the field tension meter

two point eight three four

Yes, you. Concentrate, try not to pay attention to anything else.

Yes sir. 2.834. 2.834. 2.834.

You people with children, try to soothe them, bed them down again. Mr. Hammersmith!

The wind ... Yes?

Wake up. We need your help. Oestreicher here. Star deck on the double please. A hey-rube.

But ... Right, Mr Oestreicher. On the way.

As the first officer's powerful personality took hold, the raging storm of emotion and dream subsided gradually to a sort of sullen background sea of fear, marked with fleeting whitecaps of hysteria, and Arpe found himself able to think his own thoughts again. There was no doubt about it: everyone on board the *Flyaway II* had become suddenly and totally telepathic.

But what could be the cause? It couldn't be the field. Not only was there nothing in the theory to account for it, but the field had already been effective for nearly an hour, at this same intensity, without producing any such pandemonium.

'My conclusion also,' Oestreicher said, as Arpe came on to the bridge. 'Also you'll notice that we can now see out of the ship, and that the outside sensing instruments are registering again. Neither of those things was true up to a few minutes ago; we went blind as soon as the threshold was crossed.'

'Then what's the alternative?' Arpe said. He found that it helped to speak aloud; it diverted him from the undercurrent of the intimate thoughts of everyone else. 'It must be characteristic of the space we're in, then, wherever that is. Any clues?'

'There's a sun outside,' Stauffer said, 'and it has planets. I'll have the figures for you in a minute. This I can say right away, though: it isn't Alpha Centauri. Too dim.'

Somehow, Arpe hadn't expected it to be. Alpha Centauri was in normal space, and this was obviously anything but normal. He caught the figures as they surfaced in Stauffer's mind: Diameter of primary – about a thousand miles (could that possibly be right? Yes, it was correct. But incredible). Number of planets – six. Diameter of outermost planet – about a thousand miles; distance from primary – about 50 million miles.

'What kind of a screwy system is this?' Stauffer protested. 'Six planets inside six astronomical units, and the outermost one as big as its sun? It's dynamically impossible.'

It certainly was, and yet it was naggingly familiar. Gradually the truth began to dawn on him; there was only one kind of system in which both primary and planet were consistently 1/50,000th of the distance of the outermost orbit. He suppressed it temporarily, partly to see whether or not it was possible to conceal a thought from the others under these circumstances.

'Check the orbital distances, Mr Stauffer. There should be only two figures involved.'

'Two, sir? For six planets?'

'Yes. You'll find two of the bodies occupying the same distance, and the other four at the fifty million mile distance.'

'Great Scott,' Oestreicher said. 'Don't tell me we've gotten ourselves inside an atom, sir!'

'Looks like it. Tell me, Mr Oestreicher, did you get that from my mind, or derive it from what I said?'

'I doped it out,' Oestreicher said, puzzled.

'Good; now we know something else: it *is* possible to suppress a thought in this medium. I've been holding the thought "carbon atom" just below the level of my active consciousness for several minutes.'

Oestreicher frowned, and thought: *That's good to know, it increases the possibility of controlling panic and* ... Slowly, like a sinking ship, the rest of the thought went under. The first officer was practising.

'You're right about the planets, sir,' Stauffer reported. 'I suppose this means that they'll turn out to be the same size, and that there'll be no ecliptic, either.'

'Necessarily. They're electrons. That "sun" is the nucleus.'

'But how did it happen?' Oestreicher demanded.

'I can only guess. The field gives us negative mass. We've never encountered negative mass in nature anywhere but in the microcosm. Evidently that's the only realm where it *can* exist – ergo, as soon as we attained negative mass, we were collapsed into the microcosm.'

'Great,' Oestreicher grunted. 'Can we get out, sir?'

'I don't know. Positive mass is allowable in the microcosm, so if we turned off the field, we might just keep right on staying here. We'll have to study it out. What interests me more right now is this telepathy; there must be some rationale for it.'

He thought about it. Until now, he had never believed in telepathy at all; its reported behaviour in the macrocosm had been so contrary to all known physical laws that it had been easier to assume that it didn't exist. But the laws of the macrocosm didn't apply down here; this was the domain of quantum mechanics – though telepathy didn't obey that scholium either. Was it possible that the 'para-psychological' fields were a part of the fine structure of this universe, as the electromagnetic fields of this universe itself were the fine structure of the macrocosm? If so, any telepathic effects that turned up in the macrocosm would be traces only, a leakage of residuum, fleeting and wayward, beyond all hope of control . . .

Oestreicher, he noticed, was following his reasoning with considerable interest. 'I'm not used to thinking of electrons as having any fine structure,' he said.

'Well, all the atomic particles have spin, and to measure that, you have to have some kind of a point *on* the particle being translated from one position in space to another – at least by analogy. I would say that the analogy's established now; all we have to do is look out the port.'

'You mean we might land on one of those things, sir?' Stauffer asked.

'I should think so,' Arpe said, 'if we think there's something to be gained by it. I'll leave that up to Mr Oestreicher.'

'Why not?' Oestreicher said, adding, to Arpe's surprise, 'the research chance alone oughtn't to be passed up.'

Suddenly, the background of fear, which Arpe had more and more become able to ignore, began to swell ominously; huge combers of pure panic were beginning to race over it. 'Oof,' Oestreicher said. 'We weren't covering enough – we forgot that they could pick up every unguarded word we said. And they don't like the idea.'

They didn't. Individual thoughts were hard to catch, but the main tenor was plain. These people had signed up to go to Centaurus, and that was where they wanted to go. The good possibility that they were trapped on the atomic size level was terrifying enough, but taking the further risk of landing on an electron—

Abruptly Arpe felt, almost without any words to go with it, the raw strength of Hammersmith throwing itself Canute-like against the tide. The explorer's mind had not been in evidence at all since the first shock; evidently he had quickly discovered for himself the trick of masking. For a moment the sheer militancy of Hammersmith's counter-stroke seemed to have a calming effect—

One thread of pure terror lifted above the mass. It was Celia Gospardi; she had just awakened, and her shell of bravado had been stripped completely. Following that soundless scream, the combers of panic became higher, more rapid—

'We'll have to do something about that woman,' Oestreicher said tensely. Arpe noted with interest that he was masking the thought he was speaking, quite a difficult technical trick; he tried to mask it also in the reception. 'She's going to throw the whole ship into an uproar. You were talking to her at some length last night, sir; maybe you'd better try.'

'All right,' Arpe said reluctantly, taking a step towards the door. 'I gather she's still in her—'

Flup!

Celia Gospardi *was* in her stateroom.

So was Captain Arpe.

She stifled a small vocal scream as she recognized him. 'Don't be alarmed,' he said quickly, though he was almost as alarmed as she was. 'Listen. Mr Oestreicher and everybody else: be careful about making any sudden movements with some definite destination in mind. You're likely to arrive there without having crossed the intervening distance. It's a characteristic of the space we're in.'

I read you, sir. So teleportation is an energy level jump? That could be nasty, all right.

'It's – nice of you to try to – quiet me,' the girl said timidly. Arpe noticed covertly that she could not mask worth a damn. He would have to be careful in what he said, for she would effectively make every word known throughout the ship. It was too bad, in a way. Attractive as she was in her public role, she was downright beautiful when frightened.

'Please do try to keep a hold on yourself, Miss Gospardi,' he said. 'There really doesn't seem to be any immediate danger. The ship is sound and her mechanisms are all operating as they should. We have supplies for a full year, and unlimited power; we ought to be able to get away. There's nothing to be frightened about.'

'I can't help it,' she said desperately. 'I can't even think straight. My thoughts keep getting all mixed up with everybody else's.'

'We're all having that trouble to some extent,' Arpe said. 'If you concentrate, you'll find that you can filter the other thoughts out about ninety per cent. And you'll have to try, because if you remain frightened you'll panic other people – especially the children. They're defenceless against adult emotions even *without* telepathy.'

'I – I'll try.'

'Good for you.' With a slight smile, he added, 'After all, if you think as little of your fifth husband as you say, you should welcome a little delay en route.'

It was entirely the wrong thing to say. At once, way down at the bottom of her mind, a voice cried out in soundless anguish: *But I love him!*

Tears were running down her cheeks. Helplessly, Arpe left.

He walked carefully, in no hurry to repeat the unnerving teleportation jump. In the main companionway he was waylaid by a junior officer almost at once.

'Excuse me, sir. I have a report here from the ship's surgeon. Dr Hoyle said it might be urgent and that I'd better bring it to you personally.'

'Oh. All right, what is it?'

'Dr Hoyle's compliments, sir, and he suggests that oxygen tension be checked. He has an acute surgical emergency – a passenger – which suggests that we may be running close to nine thousand.'

Arpe tried to think about this, but it did not convey very much to him, and what it did convey was confusing. He knew that spaceships, following a tradition laid down long ago in atmospheric flight, customarily expressed oxygen tension in terms of feet of altitude on Earth; but 9,000 feet – though it would doubtless cause some discomfort – did not seem to represent a dangerously low concentration. And he could see no connection at all between a slightly depleted oxygen level and an acute surgical emergency. Besides, he was too flustered over Celia Gospardi.

The interview had not ended at all the way he had hoped. But perhaps it was better to have left her grief-stricken than panic-stricken. Of course, if she broadcast her grief all over the ship, there were plenty of other people to receive it, people who had causes for grief as real as hers.

'Grief inactivates,' Oestreicher said as Arpe re-entered the bridge. 'Even at its worst, it doesn't create riots. Cheer up, sir. I couldn't have done any better, I'm sure of that.'

'Thank you, Mr Oestreicher,' Arpe said, flushing. Evidently he had forgotten to mask; 'thinking out loud' was more than a cliché down here. To cover, he proffered Hoyle's confusing message.

'Oh?' Oestreicher strode to the mixing board and scanned the big bourdon gauges with a single sweeping glance. 'He's right. We're pushing eighty-seven hundred right now. Once we cross ten thousand we'll have to order everybody into masks. I *thought* I was feeling a little light-headed. Mr Stauffer, order an increase in pressure, and get the bubble crew going, on the double.'

'Right.' Stauffer shot out.

'Mr Oestreicher, what's this all about?'

'We've sprung a major leak, sir – or, more likely, quite a few major leaks. We've got to find out where all this air is going. We may have killed Hoyle's patient already.'

Arpe groaned. Surprisingly, Oestreicher grinned.

'Everything leaks,' he said in a conversational tone. 'That's the first law of space. On the Mars run, when we disliked a captain, we used to wish him an interesting trip. This one is interesting.'

'You're a psychologist, Mr Oestreicher,' Arpe said, but he managed to grin back. 'Very well; what's the programme now? I feel some weight.'

'We were making a rocket approach to the nearest electron, sir, and we seem to be moving. I see no reason why we should suspend that. Evidently the Third Law of Motion isn't invalid down here.'

'Which is a break,' Stauffer said gloomily from the door. 'I've got the bubble crew moving, Mr Oestreicher, but it'll take a while. Captain, what are we seeing by? Gamma waves? Space itself doesn't seem to be dark here.'

'Gamma waves are too long,' Arpe said. 'Probably de Broglie waves. The illuminated sky is probably a demonstration of Obler's Paradox: it's how *our* space would look if the stars were evenly scattered throughout. That makes me think we must be inside a fairly large body of matter. And the nearest one was SV-One.'

'Oh-ho,' Stauffer said. 'And what happens to us when a cosmic ray primer comes charging through here and disrupts our atom?'

Arpe smiled. 'You've got the answer to that already. Have you detected any motion in this electron we're approaching?'

'Not much – just normal planetary motion. About fourteen miles a second – expectable for the orbit.'

'Which wouldn't be expectable at all unless we were living on an enormously accelerated time scale. By our home time scale we haven't been here a billionth of a second yet. We could spend the rest of our lives here without seeing a free neutron or a cosmic primary.'

'That's a relief,' Stauffer said; but he sounded a little dubious.

They fell silent as the little world grew gradually in the ports. There was no visible surface detail on it, and the albedo was high. As they came closer, the reasons for both effects became evident, for with each passing moment the outlines of the body grew fuzzier. It seemed to be imbedded in a sort of thick haze.

'Close enough,' Oestreicher ruled. 'We can't land the *Fly-away* anyhow; we'll have to put a couple of people off in a tender. Any suggestions, sir?'

'I'm going,' Arpe said immediately. 'I wouldn't miss an opportunity like this for anything.'

'Can't blame you, sir,' Oestreicher said. 'But that body

doesn't look like it has any solid core. What if you just sank right through to the centre?'

'That's not likely,' Arpe said. 'I've got a small increment of negative mass, and I'll retain it by picking up the ship's field with an antenna. The electron's light, but what mass it has is positive; in other words, it will repel me slightly. I won't sink far.'

'Well then, who's to go with you?' Oestreicher said, masking every word with great care. 'One trained observer should be enough, but you'll need an anchor man. I'm astonished that we haven't heard from Hammersmith already – have you noticed how tightly he shut down as soon as this subject came up?'

'So he did,' Arpe said, baffled. 'I haven't heard a peep out of him for the last hour. Well, that's his problem; maybe he had enough after Titan.'

'How about Miss Gospardi?' Stauffer suggested. 'It seems to reassure her to be with you, Captain, and it'll give her something new to think about. And it'll take an incipient panic centre out of the ship long enough to let the other people calm down.'

'Good enough,' Arpe said. 'Mr Stauffer, order the gig broken out.'

III

The little world had a solid surface, after all, though it blended so gradually into the glittering haze of its atmosphere that it was very hard to see. Arpe and the girl seemed to be walking waist-deep in some swirling, opalescent substance that was bearing a colloidal metallic dust, like minute sequins. The faint repulsions against their space suits could not be felt as such; it seemed instead that they were walking in a gravitational field about a tenth that of the Earth.

'It's terribly quiet,' Celia said.

The suit radios, Arpe noted, were not working. Luckily,

the thought-carrying properties of the medium around them were unchanged.

'I'm not at all sure that this stuff would carry sound,' he answered. 'It isn't a gas as we know it, anyhow. It's simply a manifestation of indefiniteness. The electron never knows exactly where it is; it just trails off at its boundaries into not being anywhere in particular.'

'Well, it's eerie. How long do we have to stay here?'

'Not long. I just want to get some idea of what it's like.'

He bent over. The surface, he saw, was covered with fine detail, though again he was unable to make much sense of it. Here and there he saw tiny, crooked rills of some brilliantly shiny substance, rather like mercury, and – yes, there was an irregular puddle of it, and it showed a definite meniscus. When he pushed his finger into it, the puddle dented deeply, but it did not break and wet his glove. Its surface tension must be enormous; he wondered if it were made entirely of identical subfundamental particles. The whole globe seemed to be covered by a network of these shiny threads.

Now that his eyes were becoming acclimatized, he saw that the 'air', too, was full of these shining veins, making it look distinctly marbled. The veins offered no impediment to their walking; somehow, there never seemed to be any in their immediate vicinity, though there were always many of them just ahead. As the two moved, their progress seemed to be accompanied by vagrant, small emotional currents, without visible cause or source, too fugitive to identify.

'What is that silvery stuff?' Celia demanded fearfully.

'Celia, I haven't the faintest idea. What kind of particle could possibly be sub-microscopic to an electron? It'd take a century of research right here on the spot to work up even an educated guess. This is all strange and new, utterly outside any experience man has ever had. I doubt that any words exist to describe it accurately.'

The ground, too, seemed to vary in colour. In the weak light it was hard to tell what the colours were. The variations appeared as shades of grey, with a bluish or greenish tinge here and there.

The emotional waves became a little stronger, and suddenly Arpe recognized the dominant one.

It was pain.

On a hunch, he turned suddenly and looked behind him. A twin set of broad black bootprints, as solid and sharply defined as if they had been painted, were marked out on the coloured patches.

'I don't like the look of that,' he said. 'Our ship itself is almost a planetary mass in this system, and we're far too big for this planet. How do we know what all this fine detail means? But we're destroying it wherever we step, all the same. Forests, cities, the cells of some organism, something unguessable – we've got to go back right now.'

'Believe me, I'm willing,' the girl said.

The oldest footprints, those that they had made getting out of the tender, were beginning to grow silvery at the edges, as though with hoar-frost, or with whatever fungus might attack a shadow. Or was it seepage of the same substance that made up the rills? Conjecture multiplied endlessly without answer here. Arpe hated to think of the long oval blot the tender itself would leave behind on the landscape. He could only hope that the damage would be self-repairing; there was something about this place that was peculiarly ... organic.

He lifted the tender quickly and took it out of the opalescent atmosphere with a minimum of ceremony, casting ahead for guidance to pick up the multifarious murmur of the minds on board the *Flyaway II*.

Only when he noticed that he was searching the sky visually for the ship did he realize that he was not getting anything.

'Celia? You can hear me all right telepathically, can't you?'

'Clear as a bell. It makes me feel much better, Captain.'

'Then what's wrong with the ship? I don't pick up a soul.'

She frowned. 'Why, neither do I. Where—'

Arpe pointed ahead. 'There she is, right where we left her. We could hear them all well enough at this distance when we were on the way down. Why can't we now?'

He gunned the tender, all caution forgotten. His arrival in the *Flyaway II*'s airlock was noisy, and he lost several minutes jockeying the little boat into proper seal. They both fell out of it in an inelegant scramble.

There was nobody on board the *Flyaway II*. Nobody but themselves.

The telepathic silence left no doubt in Arpe's or Celia's mind, but they searched the huge vessel thoroughly to make sure. It was deserted.

'Captain!' Celia cried. Her panic was coming back full force. 'What happened? Where could they have gone? There isn't any place—'

'I know there isn't. I don't know. Calm down a minute, Celia, and let me think.' He sat down on a stanchion and stared blindly at the hull for a moment. Breathing the thinning air was a labour in itself; he found himself wishing they had not shucked their suits. Finally he got up and went back to the bridge, with the girl clinging desperately to his elbow.

Everything was in order. It was as if the whole ship had been deserted simultaneously in an instant. Oestreicher's pipe sat smugly in its clip by the chart board; though it was empty of any trace of the self-oxygenating mixture Oestreicher's juniors had dubbed 'Old Gunpowder', the bowl was still hot.

'It can't have happened more than half an hour ago,' he whispered. 'As if they all did a jump at once – like the one that put me into your stateroom. But where to?'

Suddenly it dawned on him. There was only one answer. Of course they had gone nowhere.

'What is it?' Celia cried. 'I can see what you're thinking, but it doesn't make sense!'

'It makes perfect sense – in *this* universe,' he said grimly. 'Celia, we're going to have to work fast, before Oestreicher makes some stab in the dark that might be irrevocable. Luckily everything's running as though the crew were still here to tend it – which in fact happens to be true – so maybe two of us will be enough to do what we have to do. But you're going to have to follow instructions fast, accurately, and without stopping for an instant to ask questions.'

'What are you going to do?'

'Shut down the field. No, don't protest, you haven't the faintest idea what that means, so you've no grounds for protest. Sit down at that board over there and watch my mind every instant. The moment I think of what you're to do next, do it. Understand?'

'No, but—'

'You understand well enough. All right, let's go.'

Rapidly he began to step down the Nernst current going into the field generators, mentally directing Celia in the delicate job of holding the fusion sphere steady against the diminished drain. Within a minute he had the field down to just above the threshold level; the servos functioned without a hitch, and so, not very much to his surprise, did those aspects of the task which were supposed to be manned at all times.

'All right, now I'm going to cut it entirely. There'll be a big backlash on your board. See that master meter right in front of you at the head of the board? The black knob marked "Back EMF" is cued to it. When I pull this switch, the meter will kick over to some reading above the red line. At the same instant, you roll the knob down to *exactly* the same calibration. If you back it down too far, the Nernst will die and we'll have no power at all. If you don't go down far enough, the Nernst will detonate. You've got to catch it on the nose. Understand?'

'I – think so.'

'Good,' he said. He hoped it would be good. Normally the

roll off was handled wholly automatically, but by expending the energy evenly into the dying field; they did not dare to chance that here. He could only pray that Celia's first try would be fast. 'Here we go. Five seconds, four, three, two, one, *cut.*'

Celia twisted the dial.

For an instant, nothing happened. Then—

Pandemonium.

'Nernst crew chief, report! What are you doing? No orders were—'

'Captain! Miss Gospardi! Where did you spring from?' This was Oestreicher. He was standing right at Arpe's elbow.

'Stars! Stars!' Stauffer was shouting simultaneously. 'Hey, look! Stars! We're *back*!'

There was a confused noise of many people shouting in the belly of the *Flyaway II*. But in Arpe's brain there was blessed silence; the red foaming of raw thoughts by the hundreds was no more. His mind was his own again.

'Good for you, Celia,' he said. It was a sort of prayer. 'We were in time.'

'How did you do it, sir?' Oestreicher was saying. 'We couldn't figure it out. We were following your exploration of the electron from here, and suddenly the whole planet just vanished. So did the whole system. We were floating in another atom entirely. We thought we'd lost you for good.'

Arpe grinned weakly. 'Did you know that you'd left the ship behind when you jumped?'

'But – impossible, sir. It was right here all the time.'

'Yes, that too. It was exercising its privilege to be in two places at the same time. As a body with negative mass, it had some of the properties of a Dirac hole; as such, it had to be echoed somewhere else in the universe by an electron, like a sink and a source in calculus. Did you wind up in one of the shells of the second atom?'

'We did,' Stauffer said. 'We couldn't move out of it, either.'

'That's why I killed the field,' Arpe explained. 'I couldn't know what you would do under the circumstances, but I *was* pretty sure that the ship would resume its normal mass when the field went down. A mass that size, of course, can't exist in the microcosm, so the ship had to snap back. And in the macrocosm it isn't possible for a body to be in two places at the same time. So here we are, gentlemen – reunited.'

'Very good, sir,' Stauffer said; but the second officer's voice seemed to be a little deficient in hero-worship. 'But where is here?'

'Eh? Excuse me, Mr Stauffer, but don't you know?'

'No, sir,' Stauffer said. 'All I can tell you is that we're nowhere near home, and nowhere near the Centauri stars, either. We appear to be lost, sir.'

His glance flicked over to the bourdon gauges.

'Also,' he added quietly, 'we're still losing air.'

IV

The general alarm had alarmed nobody but the crew, who alone knew how rarely it was sounded. As for the bubble gang, the passengers who knew what that meant mercifully kept their mouths shut – perhaps Hammersmith had blustered them into silence – and the rest, reassured at seeing the stars again, were only amused to watch full-grown, grim-looking men stalking the corridors blowing soap bubbles into the air. After a while, the bubble gang vanished; they were working between the hulls.

Arpe was baffled and restive. 'Look here,' he said suddenly. 'This surgical emergency of Hoyle's – I'd forgotten about it, but it seems to have some bearing on this air situation. Let's—'

'He's on his way, sir,' Oestreicher said. 'I put a call on the bells for him as soon as – ah, here he is now.'

Hoyle was a plump, smooth-faced man with a pursed mouth and an expression of perpetual reproof. He looked absurd in his naval whites. He was also four times a Haber medal winner for advances in space medicine.

'It was a ruptured spleen,' he said primly. 'A dead give-away that we were losing oxygen. I was operating when I had the Captain called, or I'd have been more explicit.'

'Aha,' Oestreicher said. 'Your patient's a Negro, then.'

'A Negress – an eighteen-year-old girl, and incidentally one of the most beautiful women I've seen in many, many years.'

'What has her colour got to do with it?' Arpe demanded, feeling somewhat petulant at Oestreicher's obvious instant comprehension of the situation.

'Everything,' Hoyle said. 'Like many people of African extraction, she has sicklemia – a hereditary condition in which some of the red blood-cells take on a characteristic sickle-like shape. In Africa it was pro-survival, because sick-lemic people are not so susceptible to malaria as are people with normal erythrocytes. But it makes them less able to take air that's poor in oxygen – that was discovered back in the 1940s, during the era of unpressurized high altitude aero-plane flight. It's nothing that can't be dealt with by keeping sufficient oxygen in the ambient air, but—'

'How is she?' Arpe said.

'Dying,' Hoyle said bluntly. 'What else? I've got her in a tent but we can't keep that up for ever. I need normal pres-sure in my recovery room – or if we can't do that, get her back to Earth *fast*.'

He saluted sloppily and left. Arpe looked helplessly at Stauffer, who was taking spectra as fast as he could get them on to film, which was far from fast enough for Arpe, let alone the computer. The first attempt at orientation – Schmidt spherical films of the apparent sky, in the hope of identifying at least one constellation, however distorted – had come to nothing. Neither the computer nor any of the officers had been able to find a single meaningful relationship.

'Is it going to do us any good if we do find the Sun?' Oestreicher said. 'If we make another jump, aren't we going to face the same situation?'

'Here's S Doradus,' Stauffer announced. 'That's a beginning, anyhow. But it sure as hell isn't in any position I can recognize.'

'We're hoping to find the source of the leak,' Arpe reminded the first officer. 'But if we don't, I think I can calculate a fast jump – in-and-out-again. I hope we won't have to do it, though. It would involve shooting for a very heavy atom – heavy enough to be unstable—'

'Looking for the Sun?' a booming, unpleasantly familiar voice broke in from the bulkhead. It was Hammersmith, of course. Dogging his footsteps was Dr Hoyle, looking even more disapproving than ever.

'See here, Mr Hammersmith,' Arpe said. 'This is an emergency. You've got no business being on the bridge at all.'

'You don't seem to be getting very far with the job,' Hammersmith observed, with a disparaging glance at Stauffer. 'And it's my life as much as it's anybody else's. It's high time I gave you a hand.'

'We'll get along,' Oestreicher said, his face red. 'Your stake in the matter is no greater than any other passenger's—'

'Ah, that's not quite true,' Dr Hoyle said, almost regretfully. 'The emergency is medically about half Mr Hammersmith's.'

'Nonsense,' Arpe said sharply. 'If there's any urgency beyond what affects us all, it affects your patient primarily.'

'Yes, quite so,' Dr Hoyle said, spreading his hands. 'She is Mr Hammersmith's fiancée.'

After a moment, Arpe discovered that he was angry – not with Hammersmith, but with himself, for being stunned by the announcement. There was nothing in the least unlikely about such an engagement, and yet it had never entered his head even as a possibility. Evidently his unconscious still had prejudices he had extirpated from his conscious mind thirty-five years ago.

'Why have you been keeping it a secret?' he asked slowly.

'For Helen's protection,' Hammersmith said, with considerable bitterness. 'On Centaurus we may get a chance at a reasonable degree of privacy and acceptance. But if I'd kept her with me on the ship, she'd have been stared at and whispered over for the entire trip. She preferred to stay below.'

An ensign came in, wearing a space suit minus the helmet, and saluted clumsily. After he got the space suit arm up, he just left it there, resting his arm inside it. He looked like a small doll some child had managed to stuff inside a larger one.

'Bubble team reporting, sir,' he said. 'We were unable to find any leaks, sir.'

'You're out of your mind,' Oestreicher said sharply. 'The pressure is still dropping. There's a hole somewhere you could put your head through.'

'No, sir,' the ensign said wearily. 'There are no such holes. The entire ship is leaking. The air is going right out through the metal. The rate of loss is perfectly even, no matter where you test it.'

'Osmosis!' Arpe exclaimed.

'What do you mean, sir?' Oestreicher said.

'I'm not sure, Mr Oestreicher. But I've been wondering all along – I guess we all have – just how this whole business would affect the ship structurally. Evidently it weakened the molecular bonds of everything on board – and now we have good structural titanium behaving like a semipermeable membrane! I'll bet it's specific for oxygen, furthermore; a twenty per cent drop in pressure is just about what we're getting here.'

'What about the effect on people?' Oestreicher said.

'That's Dr Hoyle's department,' Arpe said. 'But I rather doubt that it affects living matter. That's in an opposite state of entropy. But when we get back, I want to have the ship measured. I'll bet it's several metres bigger in both length and girth than it was when it was built.'

'*If* we get back,' Oestreicher said, his brow dark.

'Is this going to put the kibosh on your drive?' Stauffer asked gloomily.

'It's going to make interstellar flight pretty expensive,' Arpe admitted. 'It looks like we'll have to junk a ship after one round trip.'

'Well, we effectively junked the *Flyaway I* after one *one-way* trip,' Oestreicher said reflectively. 'That's progress, of a sort.'

'Look here, all this jabber isn't getting us anywhere.' Hammersmith said. 'Do you want me to bail you out, or not? If not, I'd rather be with Helen than standing around listening to you.'

'What do you propose to do,' Arpe said, finding it impossible not to be frosty, 'that we aren't doing already?'

'Teach you your business,' Hammersmith said. 'I presume you've established our distance from S Doradus for a starter. Once I have that, I can use the star as a beacon, to collimate my next measurements. Then I want the use of an image amplifier, with a direct-reading microvoltmeter tied into the circuit; you ought to have such a thing, as a routine instrument.'

Stauffer pointed it out silently.

'Good.' Hammersmith sat down and began to scan the stars with the amplifier. The meter silently reported the light output of each, as minute pulses of electricity. Hammersmith watched it with a furious intensity. At last he took off his wrist chronometer and began to time the movements of the needle with the stop-watch.

'Bulls-eye,' he said suddenly.

'The Sun?' Arpe asked, unable to keep his tone from dripping with disbelief.

'No. That one is DQ Herculis – an old nova. It's a micro-variable. It varies by four hundredths of a magnitude every sixty-four seconds. Now we have two stars to fill our parameters; maybe the computer could give us the Sun from those? Let's try it, anyhow.'

Stauffer tried it. The computer had decided to be obtuse today. It did, however, narrow the region of search to a small sector of sky, containing approximately sixty stars.

'Does the Sun do something like that?' Oestreicher said. 'I knew it was a variable star in the radio frequencies, but what about visible light?'

'If we could mount an RF antenna big enough, we'd have the Sun in a moment,' Hammersmith said in a preoccupied voice. 'But with light it's more complicated . . . Um. If *that's* the Sun, we must be even farther away from it than I thought. Dr Hoyle, will you take my watch, please, and take my pulse?'

'Your pulse?' Hoyle said, startled. 'Are you feeling ill? The air is—'

'I feel fine, I've breathed thinner air than this and lived,' Hammersmith said irritably. 'Just take my pulse for a starter, then take everyone else's here and give me the average. I'd use the whole shipload if I had the time, but I don't. If none of you experts knows what I'm doing I'm not going to waste what time I've got explaining it to you now. Goddamn it, there are lives involved, remember?'

His lips thinned, Arpe nodded silently to Hoyle; he did not trust himself to speak. The physician shrugged his shoulders and began collecting pulse-rates, starting with the big explorer. After a while he had an average and passed it to Hammersmith on a slip of paper torn from his report book.

'Good,' Hammersmith said. 'Mr Stauffer, please feed this into Bessie there. Allow for a permitted range of variation of two per cent, and bleed the figure out into a hundred and six increments and decrements each; then tell me what the percentage is now. Can do?'

'Simple enough.' Stauffer programmed the tape. The computer jammed out the answer almost before the second officer had stopped typing: Stauffer handed the strip of paper over to Hammersmith.

Arpe watched with reluctant fascination. He had no idea what Hammersmith was doing, but he was beginning to believe that there was such a science as micro-astronomy after all.

Thereafter, there was a long silence while Hammersmith scanned one star after another. At last he sighed and said: 'There you are. This ninth magnitude job I'm lined up on now. That's the Sun. Incidentally we are a little closer to Alpha Centauri than we are from home – though God knows we're a long way from either.'

'How can you be sure?' Arpe said.

'I'm not sure. But I'm as sure as I can be at this distance. Pick the one you want to go on, make the jump, and I'll explain afterwards. We can't afford to kill any more time with lectures.'

'No,' Arpe said. 'I will do no such thing. I'm not going to throw away what will probably be our only chance – the ship isn't likely to stand more than one more jump – on a calculation that I don't know the rationale of.'

'And what's the alternative? Hammersmith demanded sneering slightly. 'Sit here and die of anoxia – and just sheer damn stubbornness?'

'I am the captain of this vessel,' Arpe said, flushing. 'We do not move until I get a satisfactory explanation of your pretensions. Do you understand me? That's my order; it's final.'

A few moments the two men glared at each other, stiff-necked as idols, each the god of his own pillbox-universe.

Hammersmith's eyelids dropped. All at once, he seemed too tired to care.

'You're wasting time,' he said. 'Surely it would be faster to check the spectrum.'

'Excuse me, Captain,' Stauffer said, excitedly. 'I just did that. And I think that star *is* the Sun. It's about eight hundred light years away—'

'*Eight hundred light years!*'

'Yes, sir, at least that. The spectral lines are about half missing, but all the ones that are definite enough to measure match nicely with the Sun's. I'm not so sure about the star Mr Hammersmith identifies as A Centaurus, but at the very least it's a spectroscopic double, and it *is* about fifty light years closer.'

'My God,' Arpe muttered. 'Eight hundred.'

Hammersmith looked up again, his expression curiously like that of a St Bernard whose cask of brandy has been spurned. 'Isn't that sufficient?' he said hoarsely. 'In God's name, let's get going. She's dying while we stand around here nit-picking!'

'No rationale, no jump,' Arpe said stonily. Oestreicher shot him a peculiar glance out of the corners of his eyes. In that moment, Arpe felt his painfully-accumulated status with the first officer shatter like a Prince Rupert's Drop; but he would not yield.

'Very well,' Hammersmith said gently. 'It goes like this. The Sun is a variable star. With a few exceptions, the pulses don't exceed the total average emission – the solar constant – by more than two per cent. The overall period is two hundred and seventy-three months. Inside that, there are at least sixty-three subordinate cycles. There's one of two hundred and twelve days. Another one lasts only a fraction over six and a half days – I forget the exact period, but it's one 1,250th of the main cycle, if you want to work it out on Bessie there.'

'I guessed something like that,' Arpe said. 'But what good does it do us? We have no tables for it—'

'These cycles have effects,' Hammersmith said. 'The six

and a half day cycle strongly influences the weather on
Earth, for instance. And the two hundred and twelve-day
cycle is reflected one for one *in the human pulse-rate.*'

'Oho,' Oestreicher said. 'Now I see. It's – Captain, this
means that we can *never* be lost! Not so long as the Sun is
detectable at all, whether we can identify it or not! We're
carrying the only beacon we need right in our blood!'

'Yes,' Hammersmith said. 'That's how it goes. It's better
to take an average of all the pulses available, since one man
might be too excited to give you an accurate figure. I'm that
overwrought myself. I wonder if it's patentable? No, a law
of nature I suppose; besides, too easily infringed, almost like
a patent on shaving. . . . But it's true, Mr Oestreicher. You
may go as far afield as you please, but your Sun stays in your
blood. You never really leave home.'

He lifted his head and looked at Arpe with hooded,
bloodshot eyes.

'Now can we go, please?' he said, almost in a whisper.
'And Captain – if this delay has killed Helen, you will
answer to me for it – if I have to chase you to the smallest,
most remote star that God ever made.'

Arpe swallowed. 'Mr Stauffer,' he said, 'prepare for
jump.'

'Where to, sir?' the second officer said. 'Back home – or
to destination?'

And there was the crux. After the next jump the *Flyaway
II* would not be spaceworthy any more. If they used it up
making Centaurus, they would be marooned; they would
have made their one round trip one-way. Besides . . . *your
drive is more important than anything else on board. Get
the passengers where they want to go by all means if it's
feasible, but if it isn't, the government wants that drive back
. . . Understand?*

'We contracted with the passengers to go to Centaurus,'
Arpe said, sitting down before the computer. 'That's where
we'll go.'

'Very good, sir,' Oestreicher said. They were the finest three words Arpe had ever heard in his life.

The Negro girl, exquisite even in her still and terrible coma, was first off the ship into the big ship-to-shore ferry. Hammersmith went with her, his big face contorted with anguish.

Then the massive job of evacuating everybody else began. Everyone – passengers and ship's complement alike – was wearing masks now. After the jump through the heavy cosmic-ray primary that Arpe had picked, a stripped nucleus which happened to be going toward Centaurus anyhow, the *Flyaway II* was leaking air as though she were made of something not much better than surgical gauze. She was through.

Oestreicher turned to Arpe and held out his hand. 'A great achievement, sir,' the first officer said. 'It'll be cut and dried into a routine after it's collimated – but they won't even know that back home until the radio word comes through, better than four years from now. I'm glad I was along while it was still new.'

'Thank you, Mr Oestreicher. You won't miss the Mars run?'

'They'll need interplanetary captains here too, sir.' He paused. 'I'd better go help Mr Stauffer with the exodus.'

'Right. Thank you, Mr Oestreicher.'

Then he was alone. He meant to be last off the ship; after living with Oestreicher and his staff for so long, he had come to see that traditions do not grow from nothing. After a while, however, the bulkhead lock swung heavily open, and Dr Hoyle came in.

'Skipper, you're bushed. Better knock it off.'

'No,' Arpe said in a husky voice, turning away from watching through the viewplate the flaming departure of the ferry for the green and brown planet, so wholly Earth-like except for the strange shapes of its continents, a thousand miles below. 'Hoyle, what do you think? Has she still got a chance?'

'I don't know. It will be nip and tuck. Maybe. Wilson – he was ship's surgeon on the *Flyaway I* – will pick her up as she lands. He's not young any more, but he was as good as they came; and with a surgeon it isn't age that matters, it's how frequently you operate. But . . . she was on the way out for a long time. She may be a little—'

He stopped.

'Go on,' Arpe said. 'Give it to me straight. I know I was wrong.'

'She was low on oxygen for a long time,' Hoyle said, without looking at Arpe. 'It may be that she'll be a little simple-minded when she recovers. Or it may not; there's no predicting these things. But one thing's for sure; she'll never dare go into space again. Not even back to Earth. The next slight drop in oxygen tension will kill her. I even advised against aeroplanes for her, and Wilson concurs.'

Arpe swallowed. 'Does Hammersmith know that?'

'Yes,' Hoyle said, 'he knows it. But he'll stick with her. He loves her.'

The ferry carrying the explorer and his fiancée, and Captain Willoughby's daughter and her Judy and many others, was no longer visible. Sick at heart, Arpe watched Centaurus III turn below him.

The planet was the gateway to the stars – for everyone on it but Daryon and Helen Hammersmith. The door that had closed behind them when they had boarded the ferry was for them no gateway to any place. It was only the door to a prison.

But it was also, Arpe realized suddenly, a prison which would hold a great teacher – not of the humanities, but of Humanity. Arpe, not so imprisoned, had no such thing to teach.

It was true that he knew how to do a great thing – how to travel to the stars. It was true that he had taken Celia Gospardi and the others where they had wanted to go. It was true that he was now a small sort of hero to his crew; and it was true that he – Dr Gordon Arpe, sometime laboratory recluse, sometime ersatz spaceship captain, sometime petty hero, had been kissed good-bye by a 3–V star.

But it was also over. From now on he could do no more than sit back and watch others refine the Arpe drive; the four-year communication gap between Centaurus and home would shut him out of those experiments as though he were Cro-Magnon Man – or Daryon Hammersmith. When next Arpe saw an Earth physicist, he wouldn't have the smallest chance of understanding a word the man said.

That was a prison, too; a prison Captain Gordon Arpe had fashioned himself, and then had thrown away the key.

'Beg pardon, Captain?'

'Oh. Sorry, Dr Hoyle. Didn't realize you were still here.' Arpe looked down for the last time on the green and brown planet, and drew a long breath. 'I said, "So be it." '

Beep

I

Josef Faber lowered his newspaper slightly. Finding the girl on the park bench looking his way, he smiled the agonizingly embarrassed smile of the thoroughly married nobody caught bird-watching, and ducked back into the paper again.

He was reasonably certain that he looked the part of a middle-aged, steadily employed, harmless citizen enjoying a Sunday break in the book-keeping and family routines. He was also quite certain, despite his official instructions, that it wouldn't make the slightest bit of difference if he didn't. These boy-meets-girl assignments always came off. Jo had never tackled a single one that had required him.

As a matter of fact, the newspaper, which he was supposed to be using only as a blind, interested him a good deal more than his job did. He had only barely begun to suspect the obvious ten years ago when the Service had snapped him up; now, after a decade as an agent, he was still fascinated to see how smoothly the really important situations came off. The *dangerous* situations – not boy-meets-girl.

This affair of the Black Horse Nebula, for instance. Some days ago the papers and the commentators had begun to mention reports of disturbances in that area, and Jo's practised eye had picked up the mention. Something big was cooking.

Today it had boiled over – the Black Horse Nebula had suddenly spewed ships by the hundreds, a massed armada that must have taken more than a century of effort on the part of a whole star-cluster, a production drive conducted in the strictest and most fanatical kind of secrecy—

And, of course, the Service had been on the spot in plenty of time. With three times as many ships, disposed with mathematical precision so as to enfilade the entire armada the moment it broke from the nebula. The battle had been a massacre, the attack smashed before the average citizen could even begin to figure out what it had been aimed at – and good had triumphed over evil once more.

Of course.

Furtive scuffings on the gravel drew his attention briefly. He looked at his watch, which said 14:58:03. That was the time, according to his instructions, when boy had to meet girl.

He had been given the strictest kind of orders to let nothing interfere with this meeting – the orders always issued on boy-meets-girl assignments. But, as usual, he had nothing to do but observe. The meeting was coming off on the dot, without any prodding from Jo. They always did.

Of course.

With a sigh, he folded his newspaper, smiling again at the couple – yes, it was the right man, too – and moved away, as if reluctantly. He wondered what would happen were he to pull away the false moustache, pitch the newspaper on the grass, and bound away with a joyous whoop. He suspected that the course of history would not be deflected by even a second of arc, but he was not minded to try the experiment.

The park was pleasant. The twin suns warmed the path and the greenery without any of the blasting heat which they would bring to bear later in the summer. Randolph was altogether the most comfortable planet he had visited in years. A little backward, perhaps, but restful, too.

It was also slightly over a hundred light-years away from Earth. It would be interesting to know how Service headquarters on Earth could have known in advance that boy would meet girl at a certain spot on Randolph, precisely at 14:58:03.

Or how Service headquarters could have ambushed with micro-metric precision a major interstellar fleet, with no more preparation than a few days' build up in the newspapers and video could evidence.

The press was free, on Randolph as everywhere. It reported the news it got. Any emergency concentration of Service ships in the Black Horse area, or anywhere else, would have been noticed and reported on. The Service did not forbid such reports for 'security' reasons or for any other reasons. Yet there had been nothing to report but that (a) an armada of staggering size had erupted with no real warning from the Black Horse Nebula, and that (b) the Service had been ready.

By now, it was a commonplace that the Service was always ready. It had not had a defect or a failure in well over two centuries. It had not even had a fiasco, the alarming sounding technical word by which it referred to the possibility that a boy-meets-girl assignment might not come off.

Jo hailed a hopper. Once inside he stripped himself of the moustache, the bald spot, the forehead-creases – all the make-up which had given him his mask of friendly innocuousness.

The hoppy watched the whole process in the rear-view mirror. Jo glanced up and met his eyes.

'Pardon me, mister, but I figured you didn't care if I saw you. You must be a Service man.'

'That's right. Take me to Service HQ, will you?'

'Sure enough.' The hoppy gunned his machine. It rose smoothly to the express level. 'First time I ever got close to a Service man: Didn't hardly believe it at first when I saw you taking your face off. You sure looked different.'

'Have to, sometimes,' Jo said, preoccupied.

'I'll bet. No wonder you know all about everything before it breaks. You must have a thousand faces each, your own mother wouldn't know you, eh? Don't you care if I know about your snooping around in disguise?'

Jo grinned. The grin created a tiny pulling sensation across one curve of his cheek, just next to his nose. He stripped away the overlooked bit of tissue and examined it critically.

'Of course not. Disguise is an elementary part of Service work. Anyone could guess that. We don't use it often, as a matter of fact – only on very simple assignments.'

'Oh.' The hoppy sounded slightly disappointed, as melodrama faded. He drove silently for about a minute. Then, speculatively: 'Sometimes I think the Service must have time-travel, the things they pull . . . well, here you are. Good luck, mister.'

'Thanks.'

Jo went directly to Krasna's office. Krasna was a Randolpher, Earth-trained, and answerable to the Earth office, but otherwise pretty much on his own. His heavy, muscular face wore the same expression of serene confidence that was characteristic of Service officials everywhere – even some that, technically speaking, had no faces to wear it.

'Boy meets girl,' Jo said briefly. 'On the nose and on the spot.'

'Good work, Jo. Cigarette?' Krasna pushed the box across his desk.

'Nope, not now. Like to talk to you, if you've got time.'

Krasna pushed a button, and a toadstool-like chair rose out of the floor behind Jo. 'What's on your mind?'

'Well,' Jo said carefully, 'I'm wondering why you patted me on the back just now for not doing a job.'

'You did a job.'

'I did not,' Jo said flatly. 'Boy would have met girl, whether I'd been here on Randolph or back on Earth. The course of true love always runs smooth. It has in all my boy-meets-girl cases, and it has in the boy-meets-girl cases of every other agent with whom I've compared notes.'

'Well, good,' Krasna said, smiling. 'That's the way we like to have it run. And that's the way we expect it to run. But,

Jo, we like to have somebody on the spot, somebody with a reputation for resourcefulness, just in case there's a snag. There almost never is, as you've observed. But – if there were?'

Jo snorted. 'If what you're trying to do is to establish pre-conditions for the future, any interference by a Service agent would throw the eventual result farther *off* the track. I know that much about probability.'

'And what makes you think that we're trying to set up the future?'

'It's obvious even to the hoppies on your own planet; the one that brought me here told me he thought the Service had time-travel. It's especially obvious to all the individuals and governments and entire populations that the Service has bailed out of serious messes for centuries, with never a single failure.' Jo shrugged. 'A man can be asked to safeguard only a small number of boy-meets-girl cases before he realizes, as an agent, that what the Service is safeguarding is the future children of those meetings. Ergo – the Service *knows* what those children are to be like, and has reason to want their future existence guaranteed. What other conclusion is possible?'

Krasna took out a cigarette and lit it deliberately; it was obvious that he was using the manoeuvre to cloak his response.

'None,' he admitted at last. 'We have some fore-knowledge, of course. We couldn't have made our reputation with espionage alone. But we have obvious other advantages: genetics, for instance, and operations research, the theory of games, the Dirac transmitter – it's quite an arsenal, and of course there's a good deal of prediction involved in all those things.'

'I see that,' Jo said. He shifted in his chair, formulating all he wanted to say. He changed his mind about the cigarette and helped himself to one. 'But these things don't add up to infallibility – and that's a qualitative difference, Kras. Take this affair of the Black Horse armada. The moment the armada appeared, we'll assume, Earth heard about it by Dirac, and started to assemble a counter-armada. But it takes *finite time* to bring together a concentration of ships and men, even if your message system is instantaneous.

'The Service counter-armada was *already on hand*. It had been building there for so long and with so little fuss that nobody even noticed it concentrating until a day or so before the battle. Then planets in the area began to sit up and take notice, and be uneasy about what was going to break. But not very uneasy; the Service always wins – that's been a statistical fact for centuries. *Centuries*, Kras. Good Lord, it takes almost as long as that, in straight preparation, to pull some of the tricks we've pulled! The Dirac gives us an advantage of ten to twenty-five years in really extreme cases out on the rim of the Galaxy, but no more than that.'

He realized that he had been fuming away on the cigarette until the roof of his mouth was scorched, and snubbed it out angrily. 'That's a very different thing,' he said, 'than knowing in a general way how an enemy is likely to behave, or what kind of children the Mendelian laws say a given couple should have. It means that we've some way of reading the future in minute detail. That's in flat contradiction to everything I've been taught about probability, but I have to believe what I see.'

Krasna laughed. 'That's a very able presentation,' he said. He seemed genuinely pleased. 'I think you'll remember that you were first impressed into the Service when you began to wonder why the news was always good. Fewer and fewer people wonder about that nowadays; it's become a part of their expected environment.' He stood up and ran a hand through his hair. 'Now you've carried yourself through the next stage. Congratulations, Jo. You've just been promoted!'

'I have?' Jo said incredulously. 'I came in here with the notion that I might get myself fired.'

'No. Come around to this side of the desk, Jo, and I'll play you a little history.' Krasna unfolded the desk-top to expose a small visor screen. Obediently Jo rose and went around the desk to where he could see the blank surface. 'I had a standard indoctrination tape sent up to me a week ago, in the expectation that you'd be ready to see it. Watch.'

Krasna touched the board. A small dot of light appeared in the centre of the screen and went out again. At the same time, there was a small *beep* of sound. Then the tape began to unroll and a picture clarified on the screen.

'As you suspected,' Krasna said conversationally, 'the Service is infallible. How it got that way is a story that started several centuries back . . .'

II

Dana Lje – her father had been a Hollander, her mother born in the Celebes – sat down in the chair which Captain Robin Weinbaum had indicated, crossed her legs, and waited, her blue-black hair shining under the lights.

Weinbaum eyed her quizzically. The conquerer Resident who had given the girl her entirely European name had been paid in kind, for his daughter's beauty had nothing fair and Dutch about it. To the eye of the beholder, Dana Lje seemed a particularly delicate virgin of Bali, despite her western name, clothing and assurance. The combination had already proven piquant for the millions who watched her television column and Weinbaum found it no less charming at first hand.

'As one of your most recent victims,' he said, 'I'm not sure that I'm honoured, Miss Lje. A few of my wounds are still bleeding. But I am a good deal puzzled as to why you're visiting me now. Aren't you afraid that I'll bite back?'

'I had no intention of attacking you personally, and I don't think I did,' the video columnist said seriously. 'It was just pretty plain that our intelligence had slipped badly in the Erskin affair. It was my job to say so. Obviously you were going to get hurt, since you're head of the bureau – but there was no malice in it.'

'Cold comfort,' Weinbaum said drily. 'But thank you, nevertheless.'

The Eurasian girl shrugged. 'That isn't what I came here about, anyway. Tell me, Captain Weinbaum – have you ever heard of an outfit calling itself Interstellar Information?'

Weinbaum shook his head. 'Sounds like a skip-tracing firm. Not an easy business, these days.'

'That's just what I thought when I first saw their letterhead,' Dana said. 'But the letter under it wasn't one that a private-eye outfit would write. Let me read part of it to you.'

Her slim fingers burrowed in her inside jacket pocket, and emerged again with a single sheet of paper. It was plain typewriter bond, Weinbaum noted automatically: she had brought only a copy with her, and had left the original of the letter at home. The copy, then, would be incomplete – probably seriously.

'It goes like this: "Dear Miss Lje: As a syndicated video commentator with a wide audience and heavy responsibilities, you need the best source of information available. We would like you to test our service, free of charge, in the hope of proving to you that it is superior to any other source of news on Earth. Therefore, we offer below several predictions concerning events to come in the Hercules and the so-called 'Three Ghosts' areas. If these predictions are fulfilled 100 per cent – no less – we ask you take us on as your correspondents for those areas, at rates to be agreed upon later. If the predictions are wrong in *any* respect, you need not consider us further." '

'H'm,' Weinbaum said slowly. 'They're confident cusses – and that's an odd juxtaposition. The Three Ghosts make up only a little solar system, while the Hercules area could include the entire star-cluster – or maybe even the whole constellation, which is a hell of a lot of sky. This outfit seems to be trying to tell you that it has thousands of field correspondents of its own, maybe as many as the government itself. If so, I'll guarantee that they're bragging.'

'That may well be so. But before you make up your mind, let me read you one of the two predictions.' The letter rustled in Dana Lje's hand. ' "At 03:16:10, on Year Day, 2090, the Hess type interstellar liner *Brindisi* will be attacked in the neighbourhood of the Three Ghosts system by four—" '

Weinbaum sat bolt upright in his swivel chair. 'Let me see that letter!' he said, his voice harsh with repressed alarm.

'In a moment,' the girl said, adjusting her skirt composedly. 'Evidently I was right in riding my hunch. Let me go on reading: "—by four heavily armed vessels flying the lights of the navy of Hammersmith II. The position of the liner at that time will be at coded co-ordinates 88-A-theta-88-aleph-D-and-per-se-and. It will—" '

'Miss Lje,' Weinbaum said. 'I'm sorry to interrupt you again, but what you've said already would justify me in jailing you at once, no matter how loudly your sponsors might scream. I don't know about the Interstellar Information outfit, or whether or not you did receive any such letter as the one you pretend to be quoting. But I can tell you that you've shown yourself to be in possession of information that only yours truly and four other men are supposed to know. It's already too late to tell you that everything you say may be held against you; all I can say now is, it's high time you clammed up!'

'I thought so,' she said, apparently not disturbed in the least. 'Then that liner *is* scheduled to hit those co-ordinates, and the coded time co-ordinate corresponds with the predicted Universal Time. Is it also true that the *Brindisi* will be carrying a top-secret communications device?'

'Are you deliberately trying to make me imprison you?' Weinbaum said, gritting his teeth. 'Or is this just a stunt, designed to show me that my own bureau is full of leaks?'

'It could turn into that,' Dana admitted. 'But it hasn't, yet. Robin, I've been as honest with you as I'm able to be. You've had nothing but square deals from me up to now. I wouldn't yellow-screen you, and you know it. If this unknown outfit has this information, it might easily have gotten it from where it hints that it got it: from the field.'

'Impossible.'

'Why?'

'Because the information in question hasn't even reached my *own* agents in the field yet – it couldn't possibly have leaked as far as Hammersmith II or anywhere else, let alone to the Three Ghosts system! Letters have to be carried on ships, you know that. If I were to send orders by ultrawave to my Three Ghosts agent, he'd have to wait three hundred and twenty-four years to get them. By ship, he can get them in a little over two months. These particular orders have only been under way to him five days. Even if somebody has read them on board the ship that's carrying them, they couldn't possibly be sent on to the Three Ghosts any faster than they're travelling now.'

Dana nodded her dark head. 'All right. Then what are we left with but a leak in your headquarters here?'

'What, indeed,' Weinbaum said grimly. 'You'd better tell me who signed this letter of yours.'

'The signature is J. Shelby Stevens.'

Weinbaum switched on the intercom. 'Margaret, look in the business register for an outfit called Interstellar Information and find out who owns it.'

Dana Lje said, 'Aren't you interested in the rest of the prediction?'

'You bet I am. Does it tell you the name of this communications device?'

'Yes,' Dana said.

'What is it?'

'The Dirac communicator.'

Weinbaum groaned and turned on the intercom again. 'Margaret, send in Dr Wald. Tell him to drop everything and gallop. Any luck with the other thing?'

'Yes, sir,' the intercom said. 'It's a one-man outfit, wholly owned by a J. Shelby Stevens, in Rico City. It was first registered this year.'

'Arrest him, on suspicion of espionage.'

The door swung open and Dr Wald came in, all six and a half feet of him. He was extremely blond, and looked awkward, gentle and not very intelligent.

'Thor, this young lady is our press nemesis, Dana Lje. Dana, Dr Wald is the inventor of the Dirac communicator, about which you have so damnably much information.'

'It's out *already*?' Dr Wald said, scanning the girl with grave deliberation.

'It is, and lots more – *lots* more. Dana, you're a good girl at heart, and for some reason I trust you, stupid though it is to trust anybody in this job. I should detain you until Year Day, videocasts or no videocasts. Instead, I'm just going to ask you to sit on what you've got, and I'm going to explain why.'

'Shoot.'

'I've already mentioned how slow communication is between star and star. We have to carry all our letters on ships, just as we did locally before the invention of the telegraph. The overdrive lets us beat the speed of light, but not by much of a margin over really long distances. Do you understand that?'

'Certainly,' Dana said. She appeared a bit nettled, and Weinbaum decided to give her the full dose at a more rapid pace. After all, she could be assumed to be better informed than the average layman.

'What we've needed for a long time, then,' he said, 'is some virtually instantaneous method of getting a message from somewhere to anywhere. Any time lag, no matter how small it seems at first, has a way of becoming major as longer and longer distances are involved. Sooner or later we must have this instantaneous method, or we won't be able to get messages from one system to another fast enough to hold our jurisdiction over outlying regions of space.'

'Wait a minute,' Dana said. 'I'd always understood that ultrawave is faster than light.'

'Effectively it is; physically it isn't. You don't understand that?'

She shook her dark head.

'In a nutshell,' Weinbaum said, 'ultrawave is radiation, and all radiation in free space is limited to the speed of light. The way we hype up ultrawave is to use an old application of wave-guide theory, whereby the real transmission of energy is at light speed, but an imaginary thing called phase velocity is going faster. But the gain in speed of transmission isn't large – by ultrawave, for instance, we get a message to Alpha Centauri in one year instead of nearly four. Over long distances, that's not nearly enough extra speed.'

'Can't it be speeded further?' she said, frowning.

'No. Think of the ultrawave beam between here and Centaurus III as a caterpillar. The caterpillar himself is moving quite slowly, just at the speed of light. But the pulses which pass along his body are going forward faster than he is – and if you've ever watched a caterpillar, you'll know that that's true. But there's a physical limit to the number of pulses you can travel along that caterpillar, and we've already reached that limit. We've taken phase velocity as far as it will go.

'That's why we need something faster. For a long time our relativity theories discouraged hope of anything faster – even the high phase velocity of the guided wave didn't contradict those theories; it just found a limited, mathematically imaginary loophole in them. But when Thor here began looking into the question of the velocity of propagation of a Dirac pulse, he found the answer. The communicator he developed does seem to act over long distances, *any* distance, instantaneously – and it may wind up knocking relativity into a cocked hat.'

The girl's face was a study in stunned realization. 'I'm not sure I've taken in all the technical angles,' she said. 'But if I'd had any notion of the political dynamite in this thing—'

'—you'd have kept out of my office,' Weinbaum said grimly. 'A good thing you didn't. The *Brindisi* is carrying a model of the Dirac communicator out to the periphery for a final test; the ship is supposed to get in touch with me from out there at a given Earth time, which we've calculated very elaborately to account for the residual Lorentz and Milne transformations involved in overdrive flight, and for a lot of other time-phenomena that wouldn't mean anything at all to you.

'If that signal arrives here at the given Earth time, then – aside from the havoc it will create among the theoretical physicists whom we decide to let in on it – we will really have our instant communicator, and can include all of occupied space in the same time-zone. And we'll have a terrific advantage over any law-breaker who has to resort to ultra-wave locally and to letters carried by ships over the long haul.'

'Not,' Dr Wald said sourly, 'if it's already leaked out.'

'It remains to be seen how much of it has leaked, ' Weinbaum said. 'The principle is rather esoteric, Thor, and the name of the thing alone wouldn't mean much even to a trained scientist. I gather that Dana's mysterious informant didn't go into technical details . . . or did he?'

'No,' Dana said.

'Tell the truth, Dana. I know that you're suppressing some of that letter.'

The girl started slightly. 'All right – yes, I am. But nothing technical. There's another part of the prediction that lists the number and class of ships you will send to protect the *Brindisi* – the prediction says they'll be sufficient, by the way – and I'm keeping that to myself, to see whether or not it comes true along with the rest. If it does, I think I've hired myself a correspondent.'

'If it does,' Weinbaum said, 'you've hired yourself a jail-bird. Let's see how much mind-reading J. Whatsit Stevens can do from the sub-cellar of Fort Yaphank.'

III

Weinbaum let himself into Stevens' cell, locking the door behind him and passing the keys out to the guard. He sat down heavily on the nearest stool.

Stevens smiled the weak benevolent smile of the very old, and laid his book aside on the bunk. The book, Weinbaum knew – since his office had cleared it – was only a volume of pleasant, harmless lyrics by a New Dynasty poet named Nims.

'Were our predictions correct, Captain?' Stevens said. His voice was high and musical, rather like that of a boy soprano.

Weinbaum nodded. 'You still won't tell us how you did it?'

'But I already have,' Stevens protested. 'Our intelligence network is the best in the Universe, Captain. It is superior even to your own excellent organization, as events have shown.'

'Its results are superior, that I'll grant,' Weinbaum said glumly. 'If Dana Lje had thrown your letter down her disposal chute, we would have lost the *Brindisi* and our Dirac transmitter both. Incidentally, did your original letter predict accurately the number of ships we would send?'

Stevens nodded pleasantly, his neatly trimmed white beard thrusting forward slightly as he smiled.

'I was afraid so,' Weinbaum leaned forward. 'Do you have the Dirac transmitter, Stevens?'

'Of course, Captain. How else could my correspondents report to me with the efficiency you have observed?'

'Then why don't our receivers pick up the broadcasts of

your agents? Dr Wald says it's inherent in the principle that Dirac 'casts are picked up by *all* instruments tuned to receive them, bar none. And at this stage of the game, there are so few such broadcasts being made that we'd be almost certain to detect any that weren't coming from our own operatives.'

'I decline to answer that question, if you'll excuse the impoliteness,' Stevens said, his voice quavering slightly. 'I am an old man, Captain, and this intelligence agency is my sole source of income. If I told you how we operated, we would no longer have any advantage over your own service, except for the limited freedom from secrecy which we have. I have been assured by competent lawyers that I have every right to operate a private investigation bureau, properly licensed, upon any scale that I may choose; and that I have the right to keep my methods secret, as the so-called "intellectual assets" of my firm. If you wish to use our services, well and good. We will provide them, with absolute guarantees on all information we furnish you, for an appropriate fee. But our methods are our own property.'

Robin Weinbaum smiled twistedly. 'I'm not a naïve man, Mr Stevens,' he said. 'My service is hard on *naïveté*. You know as well as I do that the government can't allow you to operate on a free-lance basis, supply top-secret information to anyone who can pay the price, or even free of charge to video columnists on a 'test' basis, even though you arrive at every jot of that information independently of espionage – which I still haven't entirely ruled out, by the way. If you can duplicate this *Brindisi* performance at will, we will have to have your services exclusively. In short, you become a hired civilian arm of my own bureau.'

'Quite,' Stevens said, returning the smile in a fatherly way. 'We anticipated that, of course. However, we have contracts with other governments to consider; Erskine, in particular. If we are to work exclusively for Earth, necessarily our price will include compensation for renouncing our other accounts.'

'Why should it? Patriotic public servants work for their government at a loss, if they can't work for it any other way.'

'I am quite aware of that. I am quite prepared to renounce my other interests. But I do require to be paid.'

'How much?' Weinbaum said, suddenly aware that his fists were clenched so tightly that they hurt.

Stevens appeared to consider, nodding his flowery white poll in senile deliberation. 'My associates would have to be consulted. Tentatively, however, a sum equal to the present appropriation of your bureau would do, pending further negotiations.'

Weinbaum shot to his feet, eyes wide. 'You old buccaneer! You know damned well that I can't spend my entire appropriation on a single civilian service! Did it ever occur to you that most of the civilian outfits working for us are on cost-plus contracts, and that our civilian executives are being paid just a credit a year, by their own choice? You're demanding nearly two thousand credits an hour from your own government, and claiming the legal protection that the government affords you at the same time, in order to let those fanatics on Erskine run up a higher bid!'

'The price is not unreasonable,' Stevens said. 'The service is worth the price.'

'That's where you're wrong! We have the discoverer of the machine working for us. For less than half the sum you're asking, we can find the application of the device that you're trading on – of that you can be damned sure.'

'A dangerous gamble, Captain.'

'Perhaps. We'll soon see!' Weinbaum glared at the placid face. 'I'm forced to tell you that you're a free man, Mr Stevens. We've been unable to show that you came by your information by an illegal method. You had classified facts in your possession, but no classified documents, and it's your privilege as a citizen to make guesses, no matter how educated.

'But we'll catch up with you sooner or later. Had you been reasonable, you might have found yourself in a very good position with us, your income as assured as any political income can be, and your person respected to the hilt. Now, however, you're subject to censorship – you have no idea how humiliating that can be, but I'm going to see to it that you find out. There'll be no more newsbeats for Dana Lje, or for anyone else. I want to see every word of copy that you file with any client outside the bureau. Every word that is of use to me will be used, and you'll be paid the statutory one cent a word for it – the same rate that the FBI pays for anonymous gossip. Everything I don't find useful will be killed without clearance. Eventually we'll have the modification of the Dirac that you're using, and when that happens, you'll be so flat broke that a pancake with a harelip could spit right over you.'

Weinbaum paused for a moment, astonished at his own fury.

Stevens' clarinet-like voice began to sound in the windowless cavity. 'Captain, I have no doubt that you can do this to me, at least incompletely. But it will prove fruitless. I will give you a prediction, at no charge. It is guaranteed, as are all our predictions. It is this: *You will never find that modification.* Eventually, I will give it to you, on my own terms, but you will never find it for yourself, nor will you force it out of me. In the meantime, not a word of copy will be filed with you; for, despite the fact that you are an arm of the government, I can well afford to wait you out.'

'Bluster,' Weinbaum said.

'Fact. Yours is the bluster – loud talk based on nothing more than a hope. I, however, *know* whereof I speak ... But let us conclude this discussion. It serves no purpose; you will need to see my points made the hard way. Thank you for giving me my freedom. We will talk again under different circumstances on – let me see; ah, yes, on June 9th of the year 2091. That year is, I believe, almost upon us.'

Stevens picked up his book again, nodding at Weinbaum, his expression harmless and kindly, his hands showing the marked tremor of *paralysis agitans*. Weinbaum moved helplessly to the door and flagged the turnkey. As the bars closed behind him, Stevens' voice called out: 'Oh, yes; and a Happy New Year, Captain.'

Weinbaum blasted his way back into his own office, at least twice as mad as the proverbial nest of hornets, and at the same time rather dismally aware of his own probable future. If Stevens' second prediction turned out to be as phenomenally accurate as his first had been, Captain Robin Weinbaum would soon be peddling a natty set of second-hand uniforms.

He glared down at Margaret Soames, his receptionist. She glared right back; she had known him too long to be intimidated.

'Anything?' he said.

'Dr Wald's waiting for you in your office. There are some field reports, and a couple of Diracs on your private tape. Any luck with the old codger?'

'That,' he said crushingly, 'is Top Secret.'

'Poof. That means that nobody still knows the answer but J. Shelby Stevens.'

He collapsed suddenly. 'You're so right. That's just what it does mean. But we'll bust him wide open sooner or later. We've *got* to.'

'You'll do it,' Margaret said. 'Anything else for me?'

'No. Tip off the clerical staff that there's a half-holiday today, then go take in a stereo or a steak or something yourself. Dr Wald and I have a few private wires to pull . . . and unless I'm sadly mistaken, a private bottle of aquavit to empty.'

'Right,' the receptionist said. 'Tie one on for me, Chief. I understand that beer is the best chaser for aquavit – I'll have some sent up.'

'If you should return after I am suitably squiffed,' Weinbaum said, feeling a little better already, 'I will kiss you for your thoughtfulness. *That* should keep you at your stereo at least twice through the third feature.'

As he went on through the door of his own office, she said demurely behind him, 'It certainly should'.

As soon as the door closed, however, his mood became abruptly almost as black as before. Despite his comparative youth – he was now only fifty-five – he had been in the service a long time, and he needed no one to tell him the possible consequences which might flow from possession by a private citizen of the Dirac communicator. If there was ever to be a Federation of Man in the Galaxy, it was within the power of J. Shelby Stevens to ruin it before it had fairly gotten started. And there seemed to be nothing at all that could be done about it.

'Hello, Thor,' he said glumly. 'Pass the bottle.'

'Hello, Robin. I gather things went badly. Tell me about it.'

Briefly, Weinbaum told him. 'And the worst of it,' he finished, 'is that Stevens himself predicts that we won't find the application of the Dirac that he's using, and that eventually we'll have to buy it at his price. Somehow I believe him – but I can't see how it's possible. If I were to tell Congress that I was going to spend my entire appropriation for a single civilian service, I'd be out on my ear within the next three sessions.'

'Perhaps that isn't his real price,' the scientist suggested. 'If he wants to barter, he'd naturally begin with a demand miles above what he actually wants.'

'Sure, sure . . . but frankly, Thor, I'd hate to give the old reprobate even a single credit if I could get out of it.' Weinbaum sighed. 'Well, let's see what's come in from the field.'

Thor Wald moved silently away from Weinbaum's desk while the officer unfolded it and set up the Dirac screen. Stacked neatly next to the ultraphone – a device Weinbaum had been thinking of, only a few days ago, as permanently outmoded – were the tapes Margaret had mentioned. He fed the first one into the Dirac and turned the main toggle to the position labelled *Start*.

Immediately the whole screen went pure white and the audio speakers emitted an almost instantly end-stopped blare of sound – a *beep* which, as Weinbaum already knew, made up a continuous spectrum from about 30 cycles per second to well above 18,000 cps. Then both the light and the noise were gone as if they had never been, and were replaced by the familiar face and voice of Weinbaum's local ops chief in Rico City.

'There's nothing unusual in the way of transmitters in Stevens' offices here,' the operative said without preamble. 'And there isn't any local Interstellar Information staff, except for one stenographer, and she's as dumb as they come. About all we could get from her is that Stevens is 'such a sweet old man'. No possibility that she's faking it; she's genuinely stupid, the kind that thinks Betelgeuse is something Indians use to darken their skins. We looked for some sort of list or code table that would give us a line on Stevens' field staff, but there was another dead end. Now we're maintaining a twenty-four-hour Dinwiddie watch on the place from a joint across the street. Orders?'

Weinbaum dictated the blank stretch of tape which followed: 'Margaret, next time you send any Dirac tapes in here, cut that damnable *beep* off them first. Tell the boys in Rico City that Stevens has been released, and that I'm proceeding for an Order In Security to tap his ultraphone and his local lines – this is one case where I'm sure we can persuade the court that tapping's necessary. Also – and be damned sure you code this – tell them to proceed with the tap immediately and to maintain it regardless of whether or not the court okays it. I'll thumbprint a Full Responsibility Confession for them. We can't afford to play patty-cake with Stevens – the potential is just too damned big. And oh, yes, Margaret, send the message by carrier, and send out general orders to everybody concerned not to use the Dirac again except when distance and time rule every other medium out. Stevens has already admitted that he can receive Dirac 'casts.'

He put down the mike and stared morosely for a moment at the beautiful Eridanean scrollwood of his desk-top. Wald coughed inquiringly and retrieved the aquavit.

'Excuse me, Robin,' he said, 'but I should think that would work both ways.'

'So should I. And yet the fact is that we've never picked up so much as a whisper from either Stevens or his agents. I can't think of any way that could be pulled, but evidently it can.'

'Well, let's rethink the problem, and see what we get,' Wald said. 'I didn't want to say so in front of the young lady, for obvious reasons – I mean Miss Lje, of course, not Margaret – but the truth is that the Dirac is essentially a simple mechanism in principle. I seriously doubt that there's any way to transmit a message from it which can't be detected – and an examination of the theory with that proviso in mind might give us something new.'

'What proviso?' Weinbaum said. Thor Wald left him behind rather often these days.

'Why, that a Dirac transmission doesn't *necessarily* go to all communicators capable of receiving it. If that's true, then the reasons why it is true should emerge from the theory.'

'I see. Okay, proceed on that line. I've been looking at Stevens' dossier while you were talking, and it's an absolute desert. Prior to the opening of the office in Rico City, there's no dope whatever on J. Shelby Stevens. The man as good as rubbed my nose in the fact that he's using a pseud when I first talked to him. I asked him what the "J" in his name stood for, and he said, "Oh, let's make it Jerome". But who the man behind the pseud *is*—'

'Is it possible that he's using his own initials?'

'No,' Weinbaum said. 'Only the dumbest ever do that, or transpose syllables, or retain any connection at all with their real names. Those are the people who are in serious emotional trouble, people who drive themselves into anonymity, but leave clues strewn all around the landscape – those clues are really a cry for help, for discovery. Of course we're working on that angle – we can't neglect anything – but J. Shelby Stevens isn't that kind of case, I'm sure.' Weinbaum stood up abruptly. 'Okay, Thor – what's first on your technical programme?'

'Well ... I suppose we'll have to start with checking the frequencies we use. We're going on Dirac's assumption – and it works very well, and always has – that a positron in motion through a crystal lattice is accompanied by de Broglie waves which are transforms of the waves of an electron in motion somewhere else in the Universe. Thus if we control the frequency and path of the positron, we control the placement of the electron – we cause it to appear, so to speak, in the circuits of a communicator somewhere else. After that, reception is just a matter of amplifying the bursts and reading the signal.'

Wald scowled and shook his blond head. 'If Stevens is getting out messages which we don't pick up, my first assumption would be that he's worked out a fine-tuning circuit that's more delicate than ours, and is more or less sneaking his messages under ours. The only way that could be done, as far as I can see at the moment, is by something really fantastic in the way of exact frequency control of his positron-gun. If so, the logical step for us is to go back to the beginning of our tests and re-run our diffractions to see if we can refine our measurements of positron frequencies.'

The scientist looked so inexpressibly gloomy as he offered this conclusion that a pall of hopelessness settled over Weinbaum in sheer sympathy. 'You don't look as if you expected that to uncover anything new.'

'I don't. You see, Robin, things are different in physics now than they used to be in the twentieth century. In those days, it was always presupposed that physics was limitless – the classic statement was made by Weyl, who said that, "It is the nature of a real thing to be inexhaustible in content.' We know now that that's not so, except in a remote, associational sort of way. Nowadays, physics is a defined and self-limited science; its scope is still prodigious, but we can no longer think of it as endless.

'This is better established in particle physics than in any other branch of the science. Half of the trouble physicists of the last century had with Euclidean geometry – and hence the reason why they had evolved so many recomplicated theories of relativity – is that it's a geometry of lines, and thus can be subdivided infinitely. When Cantor proved that there really is an infinity, at least mathematically speaking, that seemed to clinch the case for the possibility of a really infinite physical universe, too.'

Wald's eyes grew vague, and he paused to gulp down a slug of the licorice-flavoured aquavit which would have made Weinbaum's every hair stand on end.

'I remember,' Wald said, 'the man who taught me theory of sets at Princeton, many years ago. He used to say: "Cantor teaches us that there are many kinds of infinities. *There* was a crazy old man!" '

Weinbaum rescued the bottle hastily. 'So go on, Thor.'

'Oh.' Wald blinked. 'Yes. Well, what we know now is that the geometry which applies to ultimate particles, like the positron, isn't Euclidean at all. It's Pythagorean – a geometry of points, not lines. Once you've measured one of those points, and it doesn't matter what kind of quantity you're measuring, you're down as far as you can go. At that point, the Universe becomes discontinuous, and no further refinement is possible.

'And I'd say that our positron-frequency measurements have already gotten that far down. There isn't another element in the Universe denser than plutonium, yet we get the same frequency-values by diffraction through plutonium crystals that we get through osmium crystals – there's not the slightest difference. If J. Shelby Stevens is operating in terms of fractions of those values, then he's doing what an organist would call "playing in the cracks" – which is certainly something you can *think* about doing, but something that's in actuality impossible to do. *Hoop.*'

'Hoop?' Weinbaum said.

'Sorry. A hiccup only.'

'Oh. Well, maybe Stevens has rebuilt the organ?'

'If he has rebuilt the metrical frame of the Universe to accommodate a private skip-tracing firm,' Wald said firmly, 'I for one see no reason why we can't counter-check him – *hoop* – by declaring the whole cosmos null and void.'

'All right, all right,' Weinbaum said, grinning. 'I didn't mean to push your analogy right over the edge – I was just asking. But let's get to work on it anyhow. We can't just sit here and let Stevens get away with it. If this frequency angle turns out to be as hopeless as it seems, we'll try something else.'

Wald eyed the aquavit bottle owlishly. 'It's a very pretty problem,' he said. 'Have I ever sung you the song we have in Denmark called "Nat-og-Dag?"'

'*Hoop,*' Weinbaum said, to his own surprise, in a high falsetto. 'Excuse me. No. Let's hear it.'

The computer occupied an entire floor of the Security building, its seemingly identical banks laid out side by side on the floor along an advanced pathological state of Peano's 'space-filling curve'. At the current business end of the line was a master control board with a large television screen at its centre, at which Dr Wald was stationed, with Weinbaum looking, silently but anxiously, over his shoulder.

The screen itself showed a pattern which, except that it was drawn in green light against a dark grey background, strongly resembled the grain in a piece of highly polished mahogany. Photographs of similar patterns were stacked on a small table to Dr Wald's right; several had spilled over on to the floor.

'Well, there it is,' Wald sighed at length. 'And I won't struggle to keep myself from saying "I told you so". What you've had me do here, Robin, is to reconfirm about half the basic postulates of particle physics – which is why it took so long, even though it was the first project we started.' He snapped off the screen. 'There are no cracks for J. Shelby to play in. That's definite.'

'If you'd said, "That's flat," you would have made a joke,' Weinbaum said sourly. 'Look ... isn't there still a chance of error? If not on your part, Thor, then in the computer? After all, it's set up to work only with the unit charges of modern physics; mightn't we have to disconnect the banks that contain that bias before the machine will follow the fractional-charge instructions we give it?'

'Disconnect, he says,' Wald groaned, mopping his brow reflectively. 'The bias exists everywhere in the machine, my friend, because it functions everywhere on those same unit charges. It wasn't a matter of subtracting banks; we had to add one with a bias all of its own, to counter-correct the corrections of the computer would otherwise apply to the instructions. The technicians thought I was crazy. Now, five months later, I've proved it.'

Weinbaum grinned in spite of himself. 'What about the other projects?'

'All done – some time back, as a matter of fact. The staff and I checked every single Dirac tape we've received since you released J. Shelby from Yaphank, for any sign of inter-modulation, marginal signals, or anything else of that kind.

There's nothing, Robin, absolutely nothing. That's our net result, all around.'

'Which leaves us just where we started,' Weinbaum said. 'All the monitoring projects come to the same dead end; I strongly suspect that Stevens hasn't risked any further calls from his home office to his field staff, even though he seemed confident that we'd never intercept such calls – as we haven't. Even our local wire tapping hasn't turned up anything but calls by Stevens' secretary, making appointments for him with various clients, actual and potential. Any information he's selling these days he's passing on in person – and not in his office, either, because we've got bugs planted all over that and haven't heard a thing.'

'That must limit his range of operation enormously,' Wald objected.

Weinbaum nodded. 'Without a doubt – but he shows no signs of being bothered by it. He can't have sent any tips to Erskine recently, for instance, because our last tangle with that crew came out very well for us, even though we had to use the Dirac to send the orders to our squadron out there. If he overheard us, he didn't even try to pass the word. Just as he said, he's sweating us out—' Weinbaum paused. 'Wait a minute, here comes Margaret. And by the length of her stride, I'd say she's got something particularly nasty on her mind.'

'You bet I do,' Margaret Soames said vindictively. 'And it'll blow plenty of lids around here, or I miss my guess. The I.D. squad has finally pinned down J. Shelby Stevens. They did it with the voice-comparator alone.'

'How does that work?' Wald said interestedly.

'Blink microphone,' Weinbaum said impatiently. 'Isolates inflections on single, normally stressed syllables and matches themfl Standard I.D. searching technique, on a case of this kind, but it takes so long that we usually get the quarry by other means before it pays off. Well, don't stand there like a dummy, Margaret. Who is he?'

' "He",' Margaret said, 'is your sweetheart of the video waves, Miss Dana Lje.'

'They're crazy!' Wald said, staring at her.

Weinbaum came slowly out of his first shock of stunned disbelief. 'No, Thor,' he said finally. 'No, it figures. If a woman is going to go in for disguises, there are always three she can assume outside her own sex: a young boy, a fairy, and a very old man. And Dana's an actress; that's no news to us.'

'But – but why did she do it, Robin?'

'That's what we're going to find out right now. So we couldn't get the Dirac modification by ourselves, eh! Well, there are other ways of getting answers beside particle physics. Margaret, do you have a pick-up order out for that girl?'

'No,' the receptionist said. 'This is one chestnut I wanted to see you pull out for yourself. You give me the authority, and I send the order – not before.'

'Spiteful child. Send it, then, and glory in my gritted teeth. Come on, Thor – let's put the nutcracker on this chestnut.'

As they were leaving the computer, Weinbaum stopped suddenly in his tracks and began to mutter in an almost inaudible voice.

Wald said, 'What's the matter, Robin?'

'Nothing. I keep being brought up short by those predictions. What's the date?'

'M'm . . . June 9th. Why?'

'It's the exact date that "Stevens" predicted we'd meet again, damn it! Something tells me that this isn't going to be as simple as it looks.'

If Dana Lje had any idea of what she was in for – and considering the fact that she was 'J. Shelby Stevens' it had to be assumed that she did – the knowledge seemed not to make her at all fearful. She sat as composedly as ever before

Weinbaum's desk, smoking her eternal cigarette, and waited, one dimpled knee pointed directly at the bridge of the officer's nose.

'Dana,' Weinbaum said, 'this time we're going to get all the answers, and we're not going to be gentle about it. Just in case you're not aware of the fact, there are certain laws relating to giving false information to a security officer, under which we could heave you in prison for a minimum of fifteen years. By application of the statutes on using communications to defraud, plus various local laws against transvestism, pseudonymity and so on, we could probably pile up enough additional short sentences to keep you in Yaphank until you really *do* grow a beard. So I'd advise you to open up.'

'I have every intention of opening up,' Dana said. 'I know, practically word for word, how this interview is going to proceed, what information I'm going to give you, just when I'm going to give it to you – and what you're going to pay me for it. I knew it all that many months ago. So there would be no point in my holding out on you.'

'What you're saying, Miss Lje,' Thor Wald said in a resigned voice, 'is that the future is fixed, and that you can read it, in every essential detail.'

'Quite right, Dr Wald. Both those things are true.'

There was a brief silence.

'All right,' Weinbaum said grimly. 'Talk.'

'All right, Captain Weinbaum, pay me,' Dana said calmly. Weinbaum snorted.

'But I'm quite serious,' she said. 'You still don't know what I know about the Dirac communicator. I won't be forced to tell it, by threat of prison or by any other threat. You see, I know for a fact that you aren't going to send me to prison, or give me drugs, or do anything else of that kind. I know for a fact, instead, that you are going to pay me – so I'd be very foolish to say a word until you do. After all, it's quite a secret you're buying. Once I tell you what it is, you and the entire service will be able to read the future as I do, and then the information will be valueless to me.'

Weinbaum was completely speechless for a moment. Finally he said, 'Dana, you have a heart of purest brass, as well as a knee with an invisible gunsight on it. I say that I'm *not* going to give you my appropriation, regardless of what the future may or may not say about it. I'm not going to give it to you because the way my Government – and yours – runs things makes such a price impossible. Or is that really your price?'

'It's my real price . . . but it's also an alternative. Call it my second choice. My first choice, which means the price I'd settle for, comes in two parts: (a) to be taken into your service as a responsible officer; and, (b) to be married to Captain Robin Weinbaum.'

Weinbaum sailed up out of his chair. He felt as though copper-coloured flames a foot long were shooting out of each of his ears.

'Of all the—' he began. There his voice failed completely.

From behind him, where Wald was standing, came something like a large, Scandinavian-model guffaw being choked into insensibility.

Dana herself seemed to be smiling a little.

'You see,' she said, 'I don't point my best and most accurate knee at every man I meet.'

Weinbaum sat down again, slowly and carefully, 'Walk, do not run, to nearest exit,' he said. 'Women and childlike security officers first. Miss Lje, are you trying to sell me the notion that you went through this elaborate hanky-panky – beard and all – out of a burning passion for my dumpy and underpaid person?'

'Not entirely,' Dana Lje said. 'I want to be in the bureau, too, as I said. Let me confront you, though, Captain, with a fact of life that doesn't seem to have occurred to you at all. Do you accept as a fact that I can read the future in detail, and that that, to be possible at all, means that the future is fixed?'

'Since Thor seems able to accept it, I suppose I can too – provisionally.'

'There's nothing provisional about it,' Dana said firmly. 'Now, when I first came upon this – uh, this gimmick – quite a while back, one of the first things that I found out was that I was going to go through the "J. Shelby Stevens" masquerade, force myself on to the staff of the bureau, and marry you, Robin. At the time, I was both astonished and completely rebellious. I didn't want to be on the bureau staff; I liked my free-lance life as a video commentator. I didn't want to marry you, though I wouldn't have been averse to living with you for a while – say a month or so. And above all, the masquerade struck me as ridiculous.

'But the facts kept staring me in the face. I *was* going to do all those things. There were no alternatives, no fanciful "branches of time", no decision-points that might be altered to make the future change. My future, like yours, Dr Wald's, and everyone else's, was fixed. It didn't matter a snap whether or not I had a decent motive for what I was going to do; I was going to do it anyhow. Cause and effect, as I could see for myself, just don't exist. One event follows another because events are just as indestructible in space-time as matter and energy are.

'It was the bitterest of all pills. It will take me many years to swallow it completely, and you too. Dr Wald will come around a little sooner, I think. At any rate, once I was intellectually convinced that all this was so, I had to protect my own sanity. I knew that I couldn't alter what I was going to do, but the least I could do to protect myself was to supply myself with motives. Or, in other words, just plain rationalizations. That much, it seems, we're free to do; the consciousness of the observer is just along for the ride through time, and can't alter events – but it can comment, explain, invent. That's fortunate, none of us could stand going through motions which were truly free of what we think of as personal significances.

'So I supplied myself with the obvious motives. Since I was going to be married to you and couldn't get out of it, I set out to convince myself that I loved you. Now I do. Since I was going to join the bureau staff, I thought over all the advantages that it might have over video commentating, and found that they made a respectable list. Those are my motives.

'But I had no such motives at the beginning. Actually, there are never motives behind actions. All actions are fixed. What we called motives evidently are rationalizations by the helpless observing consciousness, which is intelligent enough to smell an event coming – and, since it cannot avert the event, instead cooks up reasons for wanting it to happen ...'

'Wow,' Dr Wald said, inelegantly, but with considerable force.

'Either "wow" or "balderdash" seems to be called for – I can't quite decide which,' Weinbaum agreed. 'We know that Dana is an actress, Thor, so let's not fall off the apple tree quite yet. Dana, I've been saving the *really* hard question for last. That question is: *How?* How did you arrive at this modification of the Dirac transmitter? Remember, we know your background, where we didn't know that of "J. Shelby Stevens". You're not a scientist. There were some fairly high-powered intellects among your distant relatives, but that's as close as you come.'

'I'm going to give you several answers to that question,' Dana Lje said. 'Pick the one you like best. They're all true, but they tend to contradict each other here and there.

'To begin with, you're right about my relatives, of course. If you'll check your dossier again, though, you'll discover that those so-called "distant" relatives were the last surviving members of my family besides myself. When they died, second and fourth and ninth cousins though they were, their estates reverted to me, and among their effects I found a sketch of a possible instantaneous communicator based on de Broglie-wave inversion. The material was in very rough form, and mostly beyond my comprehension, because I am, as you say, no scientist myself. But I was interested; I could see, dimly, what such a thing might be worth – and not only in money.

'My interest was fanned by two coincidences – the kind of coincidences that cause-and-effect just can't allow, but which seem to happen all the same in the world of unchangeable events. For most of my adult life, I've been in communications industries of one kind or another, mostly branches of video. I had communications equipment around me constantly, and I had coffee and doughnuts with communications engineers every day. First I picked up the jargon; then, some of the procedures; and eventually, a little real knowledge. Some of the things I learned can't be gotten any other way. Some other things are ordinarily available only to highly education people like Dr Wald here, and came to me by accident, in horse-play, between kisses, and a hundred other ways – all natural to the environment of a video network.'

Weinbaum found to his own astonishment that the 'between kisses' clause did not sit very well in his chest. He said, with unintentional brusqueness: 'What's the other coincidence?'

'A leak in your own staff.'

'Dana, you ought to have that set to music.'

'Suit yourself.'

'I can't suit myself,' Weinbaum said petulantly. 'I work for the government. Was this leak direct to you?'

'Not at first. That was why I kept insisting to you in person that there might be such a leak, and why I finally began to hint about it in public, on my programme. I was hoping that you'd be able to seal it up inside the bureau before my first rather tenuous contact with it got lost. When

I didn't succeed in provoking you into protecting yourself, I took the risk of making direct contact with the leak myself – and the first piece of secret information that came to me through it was the final point I needed to put my Dirac communicator together. When it was all assembled, it did more than just communicate. It predicted. And I can tell you why.'

Weinbaum said thoughtfully, 'I don't find this very hard to accept, so far. Pruned of the philosophy, it even makes some sense of the "J. Shelby Stevens" affair. I assumed that by letting the old gentleman become known as somebody who knew more about the Dirac transmitter than I did, and who wasn't averse to negotiating with anybody who had money, you kept the leak working through you – rather than transmitting data directly to unfriendly governments.'

'It did work out that way,' Dana said. 'But that wasn't the genesis or the purpose of the Stevens Masquerade. I've already given you the whole explanation of how that came about.'

'Well, you'd better name me that leak, before the man gets away.'

'When the price is paid, not before. It's too late to prevent a getaway, anyhow. In the meantime, Robin, I want to go on and tell you the other answer to your question about how I was able to find this particular Dirac secret, and you didn't. What answers I've given you up to now have been cause-and-effect answers, with which we're all more comfortable. But I want to impress on you that all apparent cause-and-effect relationships are accidents. There is no such thing as a cause, and no such thing as an effect. I found the secret because I found it; that event was fixed; that certain circumstances seem to explain why I found it, in the old cause-and-effect terms, is irrelevant. Similarly, with all your superior equipment and brains, you didn't find it for one reason, and one reason alone: because you didn't find it. The history of the future says you didn't.'

'I pays my money and I takes no choice, eh?' Weinbaum said ruefully.

'I'm afraid so – and I don't like it any better than you do.'

'Thor, what's your opinion of all this?'

'It's just faintly flabbergasting,' Wald said soberly. 'However, it hangs together. The deterministic universe which Miss Lje paints was a common feature of the old relativity theories, and as sheer speculation has an even longer history. I would say that in the long run, how much credence we place in the story as a whole will rest upon her method of, as she calls it, reading the future. If it is demonstrable beyond any doubt, then the rest becomes perfectly credible – philosophy and all. If it doesn't, then what remains is an admirable job of acting, plus some metaphysics which, while self-consistent, are not original with Miss Lje.'

'That sums up the case as well as if I'd coached you, Dr Wald,' Dana said. 'I'd like to point out one more thing. If I can read the future, then "J. Shelby Stevens" never had any need for a staff of field operatives, and he never needed to send a single Dirac message which you might intercept. All he needed to do was to make predictions from his readings, which he knew to be infallible; no private espionage network had to be involved.'

'I see that,' Weinbaum said drily. 'All right. Dana, let's put the proposition this way: *I* do not believe you. Much of what you say is probably true, but in totality I believe it to be false. On the other hand, if you're telling the whole truth, you certainly deserve a place on the bureau staff – it would be dangerous as hell *not* to have you with us – and the marriage is a more or less minor matter, except to you and me. You can have that with no strings attached; I don't want to be bought, any more than you would.

'So: if you tell me where the leak is, we will consider that part of the question closed. I make that condition not as a price, but because I don't want to get myself engaged to somebody who might be shot as a spy within a month.'

'Fair enough,' Dana said. 'Robin, your leak is Margaret Soames. She is an Erskine operative, and nobody's bubble-brain. She's a highly trained technician.'

'Well, I'll be damned,' Weinbaum said in astonishment. 'Then she's already flown the coop – she was the one who first told me we'd identified you. She must have taken on that job in order to hold up delivery long enough to stage an exit.'

'That's right. But you'll catch her, day after tomorrow. And you are not a hooked fish, Robin.'

There was another suppressed burble from Thor Wald.

'I accept the fate happily,' Weinbaum said, eyeing the gunsight knee. 'Now, if you tell me how you work your swami trick, and if it backs up everything you've said to the letter, as you claim, I'll see to it that you're also taken into the bureau and that all charges against you are quashed. Otherwise, I'll probably have to kiss the bride between the bars of a cell.'

Dana smiled. 'The secret is very simple. It's in the beep.'

Weinbaum's jaw dropped. 'The beep? The Dirac noise?'

'That's right. You didn't find out because you considered the beep to be just a nuisance, and ordered Miss Soames to cut it off all tapes before sending them in to you. Miss Soames, who had some inkling of what the beep meant, was more than happy to do so, leaving the reading of the beep exclusively to "J. Shelby Stevens" – who she thought was going to take on Erskine as a client.'

'Explain,' Thor Wald said, looking intense.

'Just as you assumed, every Dirac message that is sent is picked up by every receiver that is capable of detecting it. *Every* receiver – including the first one ever built, which is yours, Dr Wald, through the hundreds of thousands of them which will exist throughout the galaxy in the twenty-fourth century, to the untold millions which will exist in the thirtieth century, and so on. The Dirac beep is the simultaneous reception of *every one of the Dirac messages which have ever been sent, or ever will be sent.* Incidentally, the cardinal number of the total of those messages is a relatively small and of course finite number; it's far below really large finite numbers such as the number of electrons in the universe, even when you break each and every message down into individual "bits" and count those.'

'Of course,' Dr Wald said softly. 'Of course! But, Miss Lje ... how do you tune for an individual message? We tried fractional positron frequencies, and got nowhere.'

'I didn't even know fraction positron frequencies existed,' Dana confessed. 'No, it's simple – so simple that a lucky layman like me could arrive at it. You tune individual messages out of the beep by time-lag, nothing more. All the messages arrive at the same instant, in the smallest fraction of time that exists, something called a "chronon".'

'Yes,' Wald said. 'The time it takes one electron to move from one quantum-level to another. That's the Pythagorean point of time-measurement.'

'Thank you. Obviously no gross physical receiver can respond to a message that brief, or at least that's what I thought at first. But because there are relay and switching delays, various forms of feedback and so on in the apparatus itself, the beep arrives at the output end as a complex pulse which has been "splattered" along the time axis for a full second or more. That's an effect which you can exaggerate by recording the "splattered" beep on a high-speed tape, the same way you would record any event that you wanted to study in slow motion. Then you tune up the various failure-points in your receiver, to exaggerate one failure, minimize all the others, and use noise-suppressing techniques to cut out the background.'

Thor Wald frowned. 'You'd still have a considerable garble when you were through. You'd have to sample the messages—'

'Which is just what I did; Robin's little lecture to me about the ultrawave gave me that hint. I set myself to find out how the ultrawave channel carries so many messages at once, and I discovered that you people sample the incoming pulses every thousandth of a second and pass on one pip only when the wave deviates in a certain way from the mean. I didn't really believe it would work on the Dirac beep, but it turned out just as well: 90 per cent as intelligible as the original transmission after it came through the smearing device. I'd already got enough from the beep to put my plan in motion, of course – but now every voice message in it was available, and crystal-clear. If you select three pips every thousandth of a second, you can even pick up an intelligible transmission of music – a little razzy, but good enough to identify the instruments that are playing – and that's a very close test of any communications device.'

'There's a question of detail here that doesn't quite follow,' said Weinbaum, for whom the technical talk was becoming a little too thick to fight through. 'Dana, you say that you knew the course this conversation was going to take – yet it isn't being Dirac-recorded, nor can I see any reason why any summary of it would be sent out on the Dirac afterwards.'

'That's true, Robin. However, when I leave here, I will make such a transcast myself, on my own Dirac. Obviously I will – because I've *already* picked it up, from the beep.'

'In other words, you're going to call yourself up – months ago.'

'That's it,' Dana said. 'It's not as useful a technique as you might think at first, because it's dangerous to make such broadcasts while a situation is still developing. You can safely "phone back" details only after the given situation has gone to completion, as a chemist might put it. Once you know, however, that when you use the Dirac you're dealing with time, you can coax some very strange things out of the instrument.'

She paused and smiled. 'I have heard,' she said conversationally, 'the voice of the President of our Galaxy, in 3480, announcing the federation of the Milky Way and the Magellanic Clouds. I've heard the commander of a world-line cruiser, travelling from 8873 to 8704 along the world-line of the planet Hathshepa, which circles a star on the rim of NGC 4725, calling for help across eleven million light-years – but what kind of help he was calling for, or will be calling for, is beyond my comprehension. And many other things. When you check on me, you'll hear these things too – and you'll wonder what many of them mean.

'And you'll listen to them even more closely than I did, in the hope of finding out whether or not anyone was able to understand in time to help.'

Weinbaum and Wald looked dazed.

Her voice became a little more sombre. 'Most of the voices in the Dirac beep are like that – they're cries for help, which you can overhear decades or centuries before the senders get into trouble. You'll feel obligated to answer every one, to try to supply the help that's needed. And you'll listen to the succeeding messages and say: 'Did we – will we get there in time? Did we understand in time?''

'And in most cases you won't be sure. You'll know the future, but not what most of it means. The farther into the future you travel with the machine, the more incomprehensible the messages become, and so you're reduced to telling yourself that time will, after all, have to pass by at its own pace, before enough of the surrounding events can emerge to make those remote messages clear.

'The long run effect, as far as I can think it through, is not going to be that of omniscience – of our consciousness being extracted entirely from the time-stream and allowed to view its whole sweep from one side. Instead, the Dirac in effect simply slides the bead of consciousness forward from the present a certain distance. Whether it's five hundred or five thousand years still remains to be seen. At that point the law of diminishing returns sets in – or the noise-factor begins to overbalance the information, take your choice – and the observer is reduced to travelling in time at the same old speed. He's just a bit ahead of himself.'

'You've thought a great deal about this,' Wald said slowly. 'I dislike to think of what might have happened had some less conscientious person stumbled on the beep.'

'That wasn't in the cards,' Dana said.

In the ensuing quiet, Weinbaum felt a faint, irrational sense of let down, of something which had promised more than had been delivered – rather like the taste of fresh bread as compared to its smell, or the discovery that Thor Wald's Danish folk-song 'Nat-og-Dag' was only Cole Porter's 'Night and Day' in another language. He recognized the feeling: it was the usual emotion of the hunter when the hunt is over, the born detective's professional version of the *post coitum triste*. After looking at the smiling, supple Dana Lje a moment more, however, he was almost content.

'There's one more thing,' he said. 'I don't want to be insufferably sceptical about this – but I want to see it work. Thor, can we set up a sampling and smearing device such as Dana describes and run a test?'

'In fifteen minutes,' Dr Wald said. 'We have most of the unit in already assembled form on our big ultrawave receiver, and it shouldn't take any effort to add a high-speed tape unit to it. I'll do it right now.'

He went out. Weinbaum and Dana looked at each other for a moment, rather like strange cats. Then the security officer got up, with what he knew to be an air of somewhat grim determination, and seized his fiancée's hands, anticipating a struggle.

That first kiss was, by intention at least, mostly *pro forma*. But by the time Wald padded back into the office, the letter had been pretty thoroughly superseded by the spirit.

The scientist harrumphed and set up his burden on the desk. 'This is all there is to it,' he said, 'but I had to hunt all through the library to find a Dirac record with a beep still on it. Just a moment more while I make connections ...'

Weinbaum used the time to bring his mind back to the matter at hand, although not quite completely. Then two tape spindles began to whir like so many bees, and the end-stopped sound of the Dirac beep filled the room. Wald stopped the apparatus, reset it, and started the smearing tape very slowly in the opposite direction.

A distant babble of voices came from the speaker. As Weinbaum leaned forward tensely, one voice said clearly and loudly above the rest:

'Hello, Earth bureau. Lieutenant T. L. Matthews at Hercules Station NGC 6341, transmission date 13–22–2091. We have the last point on the orbit-curve of your dope-runners plotted, and the curve itself points to a small system about twenty-five light years from the base here; the place hasn't even got a name on our charts. Scouts show the home planet at least twice as heavily fortified as we anticipated, so we'll need another cruiser. We have a "can-do" from you in the beep for us, but we're waiting as ordered to get it in the present. NGC 6341 Matthews out.'

After the first instant of stunned amazement – for no amount of intellectual willingness to accept could have prepared him for the overwhelming fact itself – Weinbaum had grabbed a pencil and begun to write at top speed. As the voice signed out he threw the pencil down and looked excitedly at Dr Wald.

'Seven months ahead,' he said, aware that he was grinning like an idiot. 'Thor, you know the trouble we've had with that needle in the Hercules haystack! This orbit-curve trick must be something Matthews has yet to dream up – at least he hasn't come to me with it yet, and there's nothing in the situation as it stands now that would indicate a closing-time of six months for the case. The computers said it would take three more years.'

'It's new data,' Dr Wald agreed solemnly. 'Well, don't stop there, in God's name! Let's hear some more!'

Dr Wald went through the ritual, much faster this time. The speaker said:

'Nausentampen. Eddettompic. Berobsilom. Aimkakset-choc. Sanbetogmow. Datdectamset. Domatrosmin. Out.'

'My word,' Wald said. 'What's all that?'

'That's what I was talking about,' Dana Lje said. 'At least half of what you get from beep is just as incomprehensible. I suppose it's whatever has happened to the English language, thousands of years from now.'

'No, it isn't,' Weinbaum said. He had resumed writing, and was still at it, despite the comparative briefness of the transmission. 'Not this sample, anyhow. That, ladies and gentlemen, is code – no language consists exclusively of four syllable words, of that you can be sure. What's more, it's a version of our code. I can't break it down very far – it takes a full-time expert to read this stuff – but I get the date and some of the sense. It's March 12, 3022, and there's some kind of mass evacuation taking place. The message seems to be a routine order.'

'But why will we be using code?' Dr Wald wanted to know. 'It implies that we think somebody might overhear us – somebody else with a Dirac. That could be very messy.'

'It could indeed,' Weinbaum said. 'But we'll find out, I imagine. Give her another spin, Thor.'

'Shall I try for a picture this time?'

Weinbaum nodded. A moment later, he was looking squarely into the green-skinned face of something that looked like an animated traffic signal with a helmet on it. Though the creature had no mouth, the Dirac speaker was saying quite clearly, 'Hello, Chief. This is Thammos NGC 2287, transmission date Gor 60, 302 by my calendar, July 2, 2973 by yours. This is a lousy little planet. Everything stinks of oxygen, just like Earth. But the natives accept us and that's the important thing. We've got your genius safely born. Detailed report coming later by paw. NGC 2287 Thammos out.'

'I wish I knew my New General Catalogue better,' Weinbaum said. 'Isn't that M 41 in Canis Major, the one with the red star in the middle? And we'll be using non-humanoids there! What *was* that creature, anyhow? Never mind, spin her again.'

Dr Wald spun her again. Weinbaum, already feeling a little dizzy, had given up taking notes. That could come later, all that could come later. Now he wanted only scenes and voices, more and more scenes and voices from the future. They were better than aquavit, even with a beer chaser.

IV

The indoctrination tape ended and Krasna touched a button. The Dirac screen darkened, and folded silently back into the desk.

'They didn't see their way through to us, not by a long shot,' he said. 'They didn't see, for instance, that when one section of the government becomes nearly all-knowing – no matter how small it was to begin with – it necessarily becomes all of the government that there is. Thus the bureau turned into the Service and pushed everyone else out.

'On the other hand, those people did come to be afraid that a government with an all-knowing arm might become a rigid dictatorship. That couldn't happen and didn't happen, because the more you know, the wider your field of possible operation becomes and the more fluid and dynamic a society you need. How could a rigid society expand to other star-systems, let alone other galaxies? It couldn't be done.'

'I should think it could,' Jo said slowly. 'After all, if you know in advance what everybody is going to do—'

'But we don't, Jo. That's just a popular fiction – or, if you like, a red herring. Not all of the business of the cosmos is carried on over the Dirac, after all. The only events we can ever overhear are those which are transmitted as a message.

Do you order your lunch over the Dirac? Of course you don't. Up to now, you've never said a word over the Dirac in your life.

'And there's much more to it than that. All dictatorships are based on the proposition that government can somehow control a man's thoughts. We know now that the consciousness of the observer is the only free thing in the Universe. Wouldn't we look foolish trying to control that, when our entire physics show that it's impossible to do so? That's why the Service is in no sense a thought police. We're interested only in acts. We're an Event Police.'

'But why?' Jo said. 'If all history is fixed, why do we bother with these boy-meets-girl assignments, for instance? The meetings will happen anyhow.'

'Of course they will,' Krasna agreed immediately. 'But look, Jo. Our interests as a government depend upon the future. We operate *as if* the future is as real as the past, and so far we haven't been disappointed: the Service is 100 per cent successful. But that very success isn't without its warnings. What would happen if we *stopped* supervising events? We don't know, and we don't dare take the chance. Despite the evidence that the future is fixed, we have to take on the role of the caretaker of inevitability. We believe that nothing can possibly go wrong ... but we have to act on the philosophy that history helps only those who help themselves.

'That's why we safeguard huge numbers of courtships right through to contract, and even beyond it. We have to see to it that *every single person who is mentioned in any Dirac 'cast gets born*. Our obligation as Event Police is to make the events of the future possible, because those events are crucial to our society – even the smallest of them. It's an enormous task, believe me, and it gets bigger and bigger every day. Apparently it always will.'

'Always?' Jo said. 'What about the public? Isn't it going to smell this out sooner or later? The evidence is piling up at a terrific rate.'

'Yes and no,' Krasna said. 'Lots of people are smelling it out right now, just as you did. But the number of new people we need in the Service grows faster – it's always ahead of the number of laymen who follow the clues to the truth.'

Jo took a deep breath. 'You take all this as if it were as commonplace as boiling an egg, Kras,' he said. 'Don't you even wonder about some of the things you get from the beep? That 'cast Dana Lje picked up from Canes Venatici, for instance, the one from the ship that was travelling backwards in time? How is that possible? What could be the purpose? Is it—'

'*Pace, pace,*' Krasna said. 'I don't know and I don't care. Neither should you. That event is too far in the future for us to worry about. We can't possibly know its context yet, so there's no sense in trying to understand it. If an Englishman of around 1600 had found out about the American Revolution, he would have thought it a tragedy; an Englishman of 1950 would have a very different view of it. We're in the same spot. The messages we get from the really far future have no contexts as yet.'

'I think I see,' Jo said. 'I'll get used to it in time, I suppose, after I use the Dirac for a while. Or does my new rank authorize me to do that?'

'Yes, it does. But, Jo, first I want to pass on to you a rule of Service etiquette that must never be broken. You won't be allowed anywhere near a Dirac mike until you have it burned into your memory beyond any forgetfulness.'

'I'm listening, Kras, believe me.'

'Good. This is the rule: *The date of a Serviceman's death must never be mentioned in a Dirac 'cast.*'

Jo blinked, feeling a little chilly. The reason behind the rule was decidedly tough-minded, but its ultimate kindness was plain. He said, 'I won't forget that. I'll want that protection myself. Many thanks, Kras. What's my new assignment?'

'To begin with,' Krasna said, grinning, 'as simple a job as

I've ever given you, right here on Randolph. Skin out of here and find me that cab-driver – the one who mentioned time-travel to you. He's uncomfortably close to the truth; closer that you were in one category.

'Find him, and bring him to me. The Service is about to take in a new raw recruit!'

Beanstalk

I

The girl who came out of the Genetics Building was heroically built. From a distance, her body might have been called slim, even slight. But beside the two ugly pseudo-Greek statues which flanked the building, her height showed. She was at least eight feet tall.

She looked indecisively down the long rank of wide steps, her eyes slitted like a sleepy cat's against the morning sunlight. At the foot of the steps, a small knot of students stopped gossiping, and heads turned towards her. Sena knew the hostility in that silence.

She went down the steps, mincing over them like a dandy. They had not been laid for such a slender goddess; the risers should have been two inches higher for her, and the platforms nearly that much broader. As she approached them, the students pointedly turned their backs and examined the state of the weather.

'Damned lummoxes—' somebody muttered.

'Everybody over nine feet tall please leave the room,' said someone who had read his Carroll.

Sena had heard it all before but it still hurt. It was hard not to say, 'Out of the way, pygmies,' or in some other way make a virtue of difference.

Some of the giants had done that, in the first days; a group that had gotten notions of superiority not only to ordinary diploid human beings, but to Dr Fred himself. Their end had not been pretty, but it had been edifying. Dr Fred told that story often.

'Don't get the idea,' he said, 'that you're above your diploid fellows just because you can look down on them physically. The day may come when chromosome-doubling will be commonplace. If that day comes, it will be because the process has real advantages over normal reproduction; but those advantages are yet to be proven. If you want to see them proven, don't give yourself airs – or you won't survive to see.'

These days the giants listened to Dr Fred. He had made the giants. He was very old now, and could be expected to die before the year was out; but somehow the giants did not expect him to die. He was a man apart from the other diploids; it seemed impossible that their physical limitations could apply to him—

Careful, careful! The shortness of the diploid lifespan was not necessarily a drawback. That kind of thinking led to paranoia.

Sena passed the students, allowing herself the small pleasure of pretending that she hadn't seen them. Like most of the giants, Sena felt vaguely uneasy among them, like a parent in Toytown.

Yet it was more than that. The tallest buildings in the world were not tall enough for her, for even the tallest of buildings had entrances – entrances which would not admit a giantess unless she stooped. The whole of human civilization seemed waiting to be rebuilt, bigger and better, cleaner and higher.

And the time! The giants had so much of it! Their lifespans had not yet been measured, for, thus far, none of them had died except by violence – that had been the Pasadena pogrom, fifteen years ago. Dr Fred said that – unmolested – they should live up to six times as long as the normal diploid human. The one-in-a-thousand tetraploid adult organisms produced by nature, mostly in Lamarck's evening primrose, lived six times their normal span; and the first synthetic tetraploids had proven just as long-lived.

Of course, the very first synthetic tetraploids had been plants – *Datura*, the common chickweed, developed at

Smith College in 1937 .The U.S. Department of Agriculture had later extended the process profitably to food plants of all kinds. It was the work on rabbits and pigs conducted by Haggqvist and his associates at Stockholm's Karolinska Institute, however, which had proven artificial polyploidy possible in animals; from that momentous day in 1950, the road leading to Sena was clear.

For Sena, who was not yet forty, the whole small world was in the throes of an endless springtide; a youth that would last more than a century, with toy bridges and houses and road-planes clustered at her feet, and more than time enough to learn everything one needed to know, and the high-browed, god-like figures of lovers striding through the narrow streets of diploid man . . .

The world waited, flooded with delicate greenness that would never die.

'Sena!'

She turned. Sam Ettinger, the young, black-haired radiation expert, was running after her, traversing the cement squares in long bounds. The students scattered up the steps to watch him pass.

'Hello, Sam.'

He pulled up, smiling. His eyes crinkled at the corners; he had a way of looking at people as one might look at a sleeping tiger cub – with curiosity and admiration, yet with a certain wariness.

'You're very aloof these days,' he said. 'One would think we weren't committed to each other for this cycle.'

She reached out for his hands. 'Sam, don't. There is always so much to think about; you know that. How was the house?'-

His mouth drew down at the corners. 'No soap. When I got to the development, they had a sign up.'

' "Built to Scale"?'

'Yes. To *their* scale, as usual. The agent was willing to let me rent it if I could pay three times the tariff, but I wouldn't.'

'I don't blame you.' Sena released his hands hopelessly, all the pleasure she had taken in the sunlight oozing out of her. 'Sam, what are you going to do? Dr Fred can afford to be patient because he's old. But we've got to live in this damn society.'

'It has its drawbacks,' Sam said. 'But we can probably outlive them. Anyhow I have the outside job I told you about—'

'I still don't quite see that. I thought we were forbidden to take any part in diploid sports, by the diploids' rules.'

'We are, we are. But there's an exhibition football team of tetras, and some other exhibition teams. Strictly spectator-sports, you see. Hockey, too, and boxing. We're to play in armour, with a twenty-five pound football, against another tetra team, and the crowds come in to watch us murder each other.'

'Sam, Sam,' Sena said. She began to cry. The students watched, whispered interestedly. 'What a beastly thing to have to do – even ditch-digging would be better—'

'Ditch-digging?' Sam said quietly. 'Sena, I tried that. And I tried to get a job as a stevedore. And as a hod-carrier. And some other things of that kind. But the unions won't have it. Maybe by the time I graduate there'll be a Radiologist's Union, too!'

He looked abstractly at the bright blue sky. 'They're right, by their own lights. We're labour-saving machines. We can do more heavy work, and do it faster, than the diploids can. If the unions admitted us, sooner or later the diploids would be out of work. But this exhibition football doesn't do any economic harm to the diploids, because we aren't allowed to play against any but our own kind. Do you know what Methfessel – that's the promoter – wants to do next?'

'What?' Sena said, swallowing the lump in her throat.

'He wants to stage tournaments. The real thing; he wants to put tetras on big brewery Percherons, give them spears, swords, all the rest of the medieval armoury. If he can get government approval, he'll pay up to a hundred bucks a day.'

'For murder!'

'Not necessarily. Maurey says molybdenum steel would make a strong enough armour against a chrome-steel spear; and of course swords would be just a joke—'

'Sam, don't you see? They're making us fight each other! How long would it be before we took these tournaments seriously? Before we split up into rival groups like the Roman charioteers, with bribes and assassinations – Maurey must be mad even to consider it!'

'Well, Maurey's pretty bright,' Sam said carefully. 'Anyhow I'm not in on the tournament deal, Sena. I'm just playing this armour-plated football. It's a living. Maybe we'll even find a house after a while.'

'Maybe,' Sena said. 'In the meantime we'll just have to go on at the dorms, I suppose. I envy the ones whose parents live near the university.'

'You needn't,' Sam said. 'My folks are afraid of me. Somehow they thought the paracolchicine treatment was just going to make me turn out big and strong. Now it's "oh, not so *damned* shaggy!" You know. And my older brother hates my guts. I make him feel puny – and he claims it hurts his business connections to have a tetra in the family. He makes it sound like having a live crocodile on a leash.'

'I know,' Sena said sombrely. 'But Sam, it's worth it. There's an old Indian legend about the horned-devil caterpillar. Mother Carey offered it a drink from the Double Cup – one half held wine, and the other half held, I forget, something unpleasant; anyhow there was only one place to drink, so you got a little of each. That was how the horned devil got to be as ugly as a caterpillar, and as beautiful as a butterfly.'

Sam snorted vigorously. 'I know that business, that's just pie-in-the-sky – the old Emersonian compensation. I'm out to make things better for us poor damn caterpillars – you don't soothe me by promising me I'll be a pretty butterfly in the future. Let's go, we'll be late to Philosophy.'

The puppy was now about six weeks old: able to stagger about the laboratory floor, and to essay a tentative bounce or two, but given to frequent collapses of the rear section, and unexpected subsidences into sleep in the middle of some grandiose project. She had a box of her own, but preferred to sleep in the overturned waste-basket, which was far too small for her.

Dr Fred – Frederick R. Hyatt, on formal days – looked at her critically while she chewed on the leg of a table. Maurice St George watched them both, with an expression which seemed to indicate that he didn't know which of the two amused him more.

'But why a dog, Dr Fred?' he said. 'Surely you must have finished all the experimenting with animals before you asked for human volunteers. Something new?'

'Hmm?' Dr Fred said. 'New? No, not very. It's a line I abandoned temporarily in the early stages of the work. She's a test-tube baby; her mother was inoculated with spermatozoa in physiological salt solution, plus a dollop of paracolchicine.'

'Only the sperm chromosomes doubled, then?'

'That's right,' Dr Fred said. 'She's a triploid, not a tetraploid. Looks like she's going to be a horse just like the rest of my children, though.'

The puppy toppled over, blamed Maurey for it, spread her legs to do battle, and released a deafening yap of exasperation. Dr Fred heaved her up and put her back in her box, and threw the blanket over her head. 'Go to sleep, Decibelle.'

Finding it suddenly dark, Decibelle obediently – if involuntarily – fell asleep.

'She'll make a fine pet,' Maurey said.

'Don't you believe it, Maurice. We don't dare spring gigantic animals on the public at this stage of the game. She's going to be the world's biggest mutt – bigger than any possible Great Dane or St Bernard. We'd have an injunction slapped on us at once.'

Maurey stood up. Dr Fred noted interestedly that he did not duck his head as he did so, a gesture that was habitual with other tetraploids. Of course the normal room ceiling offered ample clearance for even the tallest of them; but the *feeling* of being boxed in was hard to battle. Maurey, evidently, had conquered it; he seemed generally to be the best adjusted to his status of all the giants, and inarguably he had the highest IQ.

Well, no reason to be surprised at that. Despite its inducing of doubling in the chromosomes – or, more accurately, its inhibition of reduction-division during mitosis – the paracolchicine process did not really have any genetic effect; that is, it did not affect the genes themselves. What it produced was called a 'mutation' because it was a change of form which bred true; but it was not a true mutation, a cataclysmic mutation springing from chemical change in the heterochromatin of the genes. Instead, it simply made it possible for the ultimate somatic expression of the individual's inheritance to come through on a tremendous scale. If there were brains in that great dark head, it was none of Dr Fred's doing.

Still, high intelligence did not imply superior ability to come to terms with one's social environment; indeed, there seemed to be some sort of rough correlation between high intelligence and the accumulation of aberrations. Dr Fred sighed inaudibly. The pioneering experiments on polyploidy hadn't had such baffling complex overtones; neither chickweed nor rabbits are much beset by emotional upsets. He wanted badly to know the nature of Maurey's adjustment; but he was not a psychologist and had no training in that field; and a lively sense of the personal inviolability of his 'children' would not allow him to require them to submit to analysis.

Through the grimy window of the lab he saw Sena and Sam, talking earnestly on the sidewalk near the building. They all *were* children, really, very alone in a settled world, and prone to whisperings, gigglings and secret societies with long names. Except that their elders might yet kill them for that secrecy – or for less understandable reasons.

'I'd like to have that dog,' Maurey said, thrusting his hands into his jacket pockets. 'I don't think I'd be afraid of the diploid neighbours.'

'Sorry, Maurice. Not yet. I need her here, anyhow.'

'I grant that you need her, since she's a triploid,' Maurey said. He looked down at the huddled blanket, frowning. 'But I want her when you're through. No matter how big she becomes, the law can't fit an ordinary dog into the description of a wild animal being kept *res naturae*, and there's no other way they could take her from me unless she gets rabid or something like that.'

Dr Fred forked his fingers and pushed his glasses up against his eyebrows. They were old glasses – for his eyesight had stabilized years ago – with battered mother-of-pearl nose-pieces, and every time he had to look down at the puppy they slid solemnly down to the end of his nose.

'You'll just make trouble,' he said. 'I've no doubt you could beat the law on such a matter, Maurice, but I don't think it advisable to try. It isn't the laws that exist now that we have to worry about, but the laws the diploids will enact later if we give them cause.'

'Forgive me, Dr Fred, but I think you're over-cautious. And perhaps overly modest, too. The giants are here to stay; and there's no point in continuing this crawfishing away from conflict with the diploids. There are plenty of provable advantages to tetraploid animals. Cats, for instance. A tetraploid cat would be the perfect answer to the rat problem. A diploid cat won't mess with a rat at all.'

'Cats,' Dr Fred said, 'have a way of lying down, grabbing, and kicking with their hind feet when they feel playful. A tetraploid cat that did that with a diploid child would rip the child to shreds.'

'Such a cat would be no more dangerous to the human child than the bathtub, statistically.'

'Maurice, you aren't dealing with statistics. You're dealing with emotions. The fifteen tetras in Pasadena had logic and reason on their side. Have you forgotten what happened to them?'

His hands still dug into his coat pockets, Maurey turned on his heel, took three quick steps, and struck the crown of his head violently against the door lintel. Dr Fred winced in sympathy. Maurey stood silently, his back to the room. Then he turned.

'No, I haven't forgotten,' he said. His eyes were streaming tears, but he wore a twisted, triumphant smile, as if his failure to sweat at his accident were a major victory. 'I won't ever forget that. The diploids are trying to forget it ever happened, but I won't ever forget. They weren't very bright, those fifteen; we've learned from them. They depended upon force alone. They forgot force's necessary adjunct.'

'And what is that, please?'

'Fraud,' Maurey said. He went out.

Dr Fred watched him go, blinking unhappily. Maybe Maurey's 'adjustment' was—

The puppy grumbled and thrust her head up out of the blanket. Dr Fred pushed her nose back down into a corner of the box. 'Sleep, dammit.'

II

The Titans, as Ira Methfessel had dubbed his first team, followed the practice plays obediently enough, but even without the armour they were slow – almost as slow in their heads as they were with their feet. Most of the young tetra-ploids had seen plenty of football games, but they had never been allowed to play during their undergraduate days; and the armour and the business of managing the shoulder jets was a double handicap.

After the first blackboard session, Ira disgustedly cut down the plays to four or five cross-bucks, to rudimentary spinners, and a few laterals – all, in essence, straight power plays. It seemed to console him somewhat to meditate that the Atlanteans were tetras, too, and unlikely to be better than the Titans; and, anyhow, all the crowd wanted was the kick of so much brute force in combat.

Ira was a diploid, the only one on the team, and the pro-moter of the whole project. He was nearly seven feet high, with flaming red hair, but he looked like a freshman among the hulking tetras. Nevertheless he treated them like high school kids. He sweated and swore, and the giants swore back cheerfully; stumbled; got in each other's way; dropped the weighted, jet-powered football, and moved their big feet a little faster. By the day of the game, Ira had whipped his bulky charges into something that resembled a flexible and resourceful team instead of a herd of rhinoceroses.

It showed results almost at once. The first ball jammed somehow and went completely out of the stadium, still under power – it was picked up later outside of town, par-tially melted – but Sam got the next one and made off with it. The Titans made their first down with a clangour like a chorus of trip-hammers.

'Good,' Ira told the huddle. 'Sam, try not to use the jets so much. If you call on them in scrimmage you'll hit somebody and we'll be penalized.'

'Okay,' Sam said. 'I was trying to get clear of that left guard of theirs—'

'Sure; but you're not allowed to hit a man while you're under power, you know that. Lucky the ref didn't see it, even though it was an accident. All right; number 80 this time.'

The Titans fanned out, the bright armour glittering and flashing in the sunlight, and crouched along the rush line. Sam took his left half spot and hunched watchfully. He was surprised to find that he was enjoying himself. The emotions of that first brutal striving, the clashing of mailed shoulders and chests, the pistoning of sollerets against the rubbery turf, released in him a willingness to hate that fifty years of Dr Fred's indoctrination had not been able quite to quell.

Hammy Saunders fired the aluminium ball back, its stripes of black and yellow paint turning lazily as it travelled. Ira snaffled it and pitched it underhand to Sam. Already the armoured figures were deploying, fending off their opponents with battering gauntlets. The crowd howled delightedly.

Sam located Hammy's red-splashed pauldrons among the scattering giants and pulled the ball back. His latticed plastic helmet was beginning to fog. An Atlantean bore down upon him.

He fired the ball. It soared, riding the thin, hazy flame of its jet. Hammy faded back, his shoulder-jets spitting, and left the turf in a tremendous leap. The ball slammed into his chest. He somersaulted once and hit the ground.

He stayed there.

The ref's whistle blew. The two teams milled and converged, rivalry forgotten. Sam wormed his way into the mêlée.

Hammy's helmet had split along the line of the front lattice. A dinged-in bar had dug out his right eye, which lay next to his ear, free upon its nerve-trunk. Blood swamped his cheek. He was still clinging to the ball. The grandstands yelled their approval.

'Look out, Sam,' Ira's voice said. 'Cripes, let a man through, you guys – he needs hospitalization – break it up, break it up–'

The giants, on both sides, made a low and ugly growling. The refs separated them hurriedly. Hammy was taken off on a stretcher.

Sam, don't you see? They're making us fight each other!

In the huddle, Ira said, 'Let's get back at 'em. They can't get away with that, fellas. Let's murder 'em. Take a straight buck over left guard – that guy hasn't had a fist in his face yet—'

'Not me,' Sam said.

'Eh? Don't give me that. Who's running this team?'

'You are,' Sam said. 'But I'm quitting. The Atlanteans didn't do anything to Hammy. It was an accident. He lost his eye because he was playing this tomfool game. I'm quitting.'

'You're yellow, you lummox.'

Sam reached out with one ring-mailed gauntlet and took Ira by his left shoulder. The tough metal bent in his grip. The diploid stumbled, flailing for balance. 'Leggo—'

'Be careful of your language, squirt,' Sam said. He swallowed. His eyes burned in their sockets, and the armour creaked across his shoulders. 'I'm sick of your games. For two cents I'd take you apart.'

He jerked his hand away suddenly. The pauldron came away with it, with a screech of outraged metal. Sam took it in both hands and crumpled it methodically, like wrapping paper. The sudden, irregular *give* of the plate in his palms was like the breaking of bone, and he was shaken by an ugly love for it.

'Here,' he said. He handed the wadded plating back to the promoter, his mouth twitching with the bitterness in it. 'Be glad it wasn't you. I'm quitting – *now* do you understand?'

Ira took the bunched mail numbly, staring at Sam through his slotted plastic fishbowl. 'Look, Sam,' he said. 'You're blaming *me* for this. I didn't do it. Nobody did it, like you say. You knew this job was dangerous, and so did Hammy—'

The ref's whistle screamed, and the whole tense group was marched five yards down the field for overtime. None of the Titans seemed to notice; they followed the quarrel and gathered around it again. Some of the Atlanteans began to filter over into the Titan huddle. The crowd grumbled with puzzled impatience.

'Ira's right,' Chris Harper said. 'It isn't his fault.'

'I said I knew it was an accident,' Sam growled. 'Just the kind of accident the shrimps in the stands came to see. The penalty we pay for being so numerous – if there were fewer of us, we could make our livings in a side-show. I'm through with both kinds!'

He strode off the field, his pads creaking. The crowd booed him enthusiastically.

Maurey was waiting for him as he came out of the locker-room. The older giant was wearing a crooked smile that puzzled Sam, and, in his present mood, infuriated him despite his respect for Maurey.

'What are you grinning at?' he demanded. 'You think it's funny when a man loses an eye – like them – back there?' He jerked a thumb over his shoulder in the direction of the bleachers.

'Not at all,' Maurey said soothingly. The crooked smile dimmed a little, but did not disappear. 'I take it pretty seriously, I assure you, Sam. Going back to the lab?'

'Yes, if you need me. Anyhow I've no place else to go until Sena is free.'

'Good,' Maurey said. 'I'll give you a lift, if you like. My roader's in the parking lot.'

Maurey said nothing further until the roadplane was on the express lane leading back toward the city; and even then he seemed only to be making talk. Finally he turned up a ski-jump, snapped the rotor open, and climbed the craft steeply. 'So Ira's finally made you mad,' he said.

'Yeah,' Sam said. The word was muffled; he sat immobile, staring straight ahead. He had already begun to feel a little guilty for his outburst on the field, but Maurey's question made him rebellious all over again. 'I think maybe it's about time this damned culture found productive jobs for us, Maurey. The accident wasn't Ira's fault. It was the fault of all the diploids.'

'All?' Maurey said quietly.

'Yes, all. I suppose you want me to except Dr Fred. I won't. He means well, but he's been one of the major factors in keeping us satisfied – moderately, anyhow – with the status quo. That can't last forever.'

Maurey cast a sidelong glance at him. 'I've been telling Dr Fred that,' he said, 'but he's too old to change. We'll have to make our future ourselves, if it's to suit us.'

'You've something in mind?' Sam said curiously.

'Yes, I think so. I want to be sure I'm not just setting up a Pasadena before I talk too much about it, though.'

'I'm a quiet sort,' Sam said. 'Can't you give me some idea—'

'Well, in essence it's quite simple. I want to start a home-steading project. The unmilitarized part of the Moon has just been declared public land; I think we could occupy it profitably.'

'That sounds unlikely,' Sam said. 'Anyhow it *is* Pasadena all over again, Maurey, and it's just what the diploids would like – get us all into a ghetto somewhere where they could bomb us to extinction all at once.'

Maurey peered downward, and then began to sidle the roader toward the Earth. 'I'm not quite so stupid, Sam,' he said, smiling again. 'Of course it's like the Pasadena thing on the surface – intentionally. I made up my mind long ago that the only way to get anything from the diploids is to seem to be doing things their way. *Seem* to be, Sam. Actually I think our tetraploid homeland won't last more than a month or so. By the end of that time we'll either be extinct, or be in a position to dictate the terms of our tenancy on diploid Earth. I hope some day we can have a *real* planet, and I rather hope it'll be this one; the tetras are bound to multiply. Most parents will become more and more reluctant to deny their children the advantage of tetraploidy in a world where tetras are part of the normal order of things.'

Sam was a little confused. 'You forget the low-fertility angle,' he objected. 'There's still plenty of religious and moralistic opposition to our plural-marriage arrangements; and even more sentimental worship of motherhood. Many a family would drop dead before it made a prospective daughter into a giantess – that's why we're so short of women – and the diploids are already too proud of being more fertile than we are.'

'Sure, sure. That's as may be. I didn't say this was going to be easy.' Maurey dropped the plane skilfully to the highway, collapsed the rotor, and guided it to the lane that led to the university. 'My point is that we'll have to seem to be playing along with the diploids for a while. In the final analysis, our job is just this: to trick the diploids into putting weapons into our hands. Dr Fred has already given us one—'

'You mean our size?'

'No, that's no great advantage yet, and besides it's not the kind of weapon I mean. Have you met Decibelle?'

'That fool puppy? You bet. She mangles my shoelaces.'

'Dr Fred doesn't see all the implications, I'm glad to say,' Maurey said. 'Also the reactionless effect that we're working on will be a weapon sooner or later; we have you to thank for that, Sam. But I'm depending most of all upon Ira and his silly tournament.'

'Great Jupiter,' Sam said. 'The next thing you'll be saying is that you want me to go back and play football for Ira.'

'I want you to do exactly that,' Maurey said calmly. 'I can't order you to do it, because I'm your superior only in the lab; but I'd appreciate it if you would. I want Ira's tournament promoted for all it's worth. If we can get the reactionless effect developed in time, I'm going to give that to

Ira, too; it'll be a great improvement over those shoulder-jets; and as a weapon it could be one of the deadliest side-arms in history – among other things. I leave it to you, Sam, to imagine what those other things might be. Here we are.'

He garaged the roadplane in the radiation lab's basement and swung the door open. 'Coming in?'

'Sure,' said Sam abstractly. 'What's the great to-do about the reactionless effect, Maurey? It's only a laboratory toy as far as I can see. I'm pretty well convinced that we'll find out where the backlash is going to before long.'

'You haven't found it yet?'

'No-o-o-o. But I think it must be getting recirculated somehow, the way a regenerative circuit uses back EMF. It isn't logical that there should be no reaction at all!'

Maurey shrugged. 'Mr Newton's Third Law of Motion may not be any more universal than any of his other laws,' he said. 'By all means try to find where the recoil is going, so there'll be no mistake. But if it turns out that there *isn't* any "equal and opposite effect"—'

'There will be,' Sam said flatly. 'There always is. Aren't you coming up, Maurey?'

'No, I've got work elsewhere. You have your key, haven't you? All right – see you tomorrow.'

Sam climbed the stairs and let himself into the lab in a brown study; he was hardly aware of Maurey's absence. Essentially a scientist, Sam was easily swayed when it came to political manoeuvring – but the faintest smell of a technical puzzle was enough to wipe politics from his mind. He had already forgotten the quarrel with Methfessel; he had almost forgotten Maurey's hints about a tetraploid 'home-land'. The suggestion that Newton's Third Law of Motion actually might not apply to his toy had been enough to enlist his total attention.

He plugged the power jacks into the apparatus and waited for the tubes to warm up. That waste of power, made necessary by the impossibility of using transistors in the apparatus, he understood; but this other thing—

The experiment, originally, had been set up to explore some side-effects of magneton-rotation; a routine high-altitude project. Maurey and Sam had guessed that the government hoped to see some sort of antigravity come out of the new Blackett-Dirac theory of magnetism. Thus far, no such thing had appeared; instead—

He touched the key experimentally. Across the room, a large bell chimed pleasantly, though it was not in any way connected with the apparatus. Sam got up, took down the bell, and put up the regular target. The machine was behaving as always. Every erg of energy that went into it was metered; even the losses in metering were figured. And the amount of thrust that that invisible pulse shot at the target always equalled exactly the amount of power that the apparatus used.

There was no equivalent 'recoil'.

Suppose that apparent lack of feedback was real, as Maurey had suggested? Suppose that, for once, an action did *not* involve an equal and opposite reaction? Suppose that, for once, an object that was pushed *didn't push back*?

Of course the target pushed back, but that was secondary – *ex post facto*, as it were. He changed the metering set-up and started again. There was no gain in the amount of heat put out by the tubes when the device was 'fired'. The wiring didn't heat up, either. On a hunch, he made a free coil of the main power pack lead, made a foray next door to liberate a breaker of liquid air from the pressure lab, and dipped the coil in it.

The target burst the moment the key was closed. Excitedly Sam checked the target meter readings against the lowered resistance of the cold coil. It checked to the last decimal place. Nothing lost in resistance, then? The boiling of the liquid air hadn't speeded visibly when the pulse was launched, but the eye couldn't be trusted to detect such a thing. As a last check, he bottled the liquid air and the coil in a Dewar flask with a sensitive Rahm transducer for a cork. The transducer he rigged to a kymograph.

He fired the device four times. The steadily rising line on the kymograph drum showed not the slightest joggle from bottom to top. And since he'd already failed repeatedly to detect any radio or suboptical effects—

Newton's Third Law of Motion was a gone goose.

A math to describe it could wait; as a matter of fact it would be fairly simple to express it as a matrix discontinuity. What interested Sam now was a way of reassembling the apparatus so that it would be portable. Even to his unpractical eye, the advantages of portability were evident. If a man could hold a thing like this in his hand, and apply just as much push to an object as there was power available – if he could, for instance, convert a couple of thousand kilowatts into physical thrust against a heavy load – two or three men might lever a heavy locomotive out of a culvert – or—

The engineering of a compact projector was not difficult. All but two of the tubes could be replaced by a couple of 6V6's without much loss in efficiency, and the loss could be expressed as heat and dissipated harmlessly by discharging the pulse from a flanged tube with a reflector behind it: the flanging might be charged, too, to make a focusing field, and the tube could be silver and act as a wave-guide—

In another hour, Sam had a thing that might have been a twenty-first-century cross-bow, without the bow. It was certainly awkward, but it worked. Sam sat in a window of the lab and knocked off the hats of passers-by until it got too dark outside to make the sport safe. Then he locked up and went back to the dorms, whistling tunelessly.

A student of history might have known where to expect the missing back-blast; but Sam was only a scientist.

*   *   *

III

The windows of the graduate lab in radiation were like the windows of every other college laboratory – big, inadequately puttied, and long unwashed. Maurey did not see the device on the bench until he had been in the lab for several minutes, for the sunlight was slanted the other way, and the main workbench was only dimly lit in the daytime unless the sun was directly upon it.

When he did see it, he drew his breath in sharply. It took only a moment to check the leads, and to confirm that this was what had been yesterday only a confusion of tubes and spaghetti. Maurey looked it over carefully. What he saw raised his estimate of Sam by quite a few notches. Yesterday the generator of the one-way-push had taken up as much space as an ancient super-heterodyne radio; now it was all neatly assembled along a single axis, scarcely more difficult to handle than a shotgun, except for the leads.

Sam had given Maurey his weapon.

Maurey shut the door quietly and locked it. The discarded, empty Dewar flask told its own story. It also suggested something that Sam evidently had overlooked. Maurey found Sam's free coil and made a receptacle for it, with clips to hold the Dewar flask under the silver barrel of the gun. He got his liquid air from the same source that Sam had; but instead of corking the flask, he soldered a conduit from it to a tiny, fan-driven booster-generator. Two flashlight batteries and a small transformer finished the job; he cast the power-lead free.

The device was completely portable now, and the first successful perpetual-motion machine in history – as long as the liquid air held out. In a heavy-duty, semi-portable projector, Maurey thought, some of the output could be diverted to run a compressor, completing the cycle. The prospect rather dazed him.

Maurey took the projector with him, bundled into an innocuous lump in old newspapers. On the campus, students waved to him; even diploids. Maurey was well-liked; his air of mild, cosmopolitan amusement made him envied by the young; and among the college students there was an idealistic 'equal rights for tetras' movement which Maurey had taken great care to further. The diploid kids loved Maurey, where they only respected the younger giants.

'Hi, Maurey. Whatcha got?'

'Wet-pack,' Maurey said. 'How are you, June? Get over that squabble with the parents all right?'

'Yes, thanks to you. Are you coming to our meeting tonight?'

'I hope so. Don't wait up for me, though.'

He tucked a flap of newspaper over the weapon, nodded to the girl, and turned down the gravel path. At the other end of the campus he saw another giant, but the distance was too great to see who it was. It was male, and that was all Maurey could determine. He had a sudden urge to run shouting toward the towering figure, to declare war at once upon all the scurrying pygmies, pick them off like clay pigeons with the invincible thing he had packed in newspapers under his arm—

Not yet. He went on, smiling to the diploid youngsters who worshipped him.

He realized that his own immediate plan was far from perfect. The most important thing had been to get the weapon out of the Radiology Building, where the chances were good that anyone who found it might understand it – or understand enough of it to become dangerously curious.

Sam wouldn't trouble too much about its disappearance. He would assume that Maurey had taken it, and as soon as he found that to be true he would be satisfied. Dr Fred, on the other hand, would 'know' immediately that the thing was not a weapon, but some toy of the radiation labs; so probably the best place for it was in Dr Fred's safe, at least until it could be delivered to Methfessel. Dr Fred was almost fanatical about respecting the property rights of the giants. When he found the projector he would identify it as Maurey's and leave it alone.

All this was parlour psychology, based insecurely upon Maurey's estimate of the people involved, but it would have to do. The most important thing was to persuade Sam not to publish his findings. It would take some doing, for Sam, even more than the ordinary graduate student, depended upon his scientific reputation for his small income. A discovery as revolutionary as this might net him an assistant-professorship.

Maurey was mildly surprised to find Dr Fred's lab empty. The old man rarely went out these days; he held the rank of Professor Emeritus, hence had no classes to teach, and spent most of his waking hours (which meant twenty out of every twenty-four) making microtome sections, fixing, staining, mounting, sketching, and filing the thousands of tissue specimens necessary to any experiment in polyploidy. Evidently he was taking one of his unpredictable four-hour naps.

Well, that was all to the good. Maurey knelt before the safe. One of the sloppy human abilities tetraploidy had sharpened for Maurey was hearing: the muscles of his middle ear were as sensitive as those of the pupils of his eyes, and could reduce the vibrating surface of his eardrum to a taut spot no bigger than a pinhead when he was listening intently. He could hear sounds from 4 cycles to 30,000, and do it selectively; his mind had recorded the tiny tinkle of tumblers almost automatically the first time Dr Fred had opened that safe.

The door swung open, and Maurey sniffed with annoyance. Here was a problem he hadn't anticipated. The safe was full of papers. More than full; it was stuffed. Two of the three top pigeon-holes were taken up by slide-boxes, the third by the familiar cardboard mailing tubes in which para-colchicine ampules came into the lab; but the rest of the safe was packed with note-paper, graph-paper, drawing-paper, photomicrographs, letters, thin pamphlets with long titles, file-cards, and what seemed to be a thousand shiny black booklets of silicone-treated lens-paper. A sizable proportion of the mess sagged chummily into Maurey's lap the moment the safe door moved back.

He swore. From a box under the table something sneezed in answer, and Decibelle stuck her nose out and regarded Maurey with reproachful brown eyes, her eyebrows going up and down independently.

'Go to sleep, pooch. Damn. What am I going to do with—'

He realized that he was talking to himself and stopped. As a preliminary measure, he took all the papers out and made four piles of them on the table. Obviously the stuff had been jammed into the safe in no particular order, so it would do no harm to redistribute it in some way which would be more economical of space. Probably the best way would be to stack according to size, all the films in one pile, all the notes in another, the publications in another, and so on, and then repack—

Maurey's hand, turning over a crumpled letter-size sheet, paused in mid-air. After a long moment it continued the arrested motion, laying the sheet with meticulous care upon the proper pile. Maurey picked up the sheet that had been under it.

CARLIN, SENA HYATT

(Jane Hyatt, Anthony Armisted Carlin)
Series O–573–9–002

Sex-linked double-diploid with marked tetraploidy; cf. chromosome charts, 2, 3, 6, 8, 9, 10, 14, 15, 18, 21, 22, 24. Heavy crossing-over on diploid chromosomes. Triploid x-chromosome. Somatically an apparent normal tetraploid individual with only slight schisming . . .

It was no news of course, that Sena had Hyatt blood in

her; most of the older giants did; only the youngest gener-
ation had had to suffer by the court order forbidding Dr
Fred to contribute germ cells to the polyploidy experiments.

But what the hell was *double-diploidy*? Twice two was
four, any fool knew that. Yet Dr Fred must have some
reason for calling Sena a 'double-diploid' instead of a tetraploid. And that reference to 'schisming' – the awkward word
was deliberate, an avoidance of 'schizoid' or any other term
that might have referred to Sena's psychology; that sentence
began with the crucial term 'somatically'.

Maurey was not a geneticist, but he knew his own background, and he was used to scientific shorthand. There was
only one interpretation possible. Some of the twenty-four
chromosome pairs which carried the human inheritance,
and which should have been given Sena in double measure,
had not doubled – had not *been* doubled, deliberately, for
the placid failure of Dr Fred's record to evince surprise be-
trayed foreplanning. Many of those that had doubled were
still acting as sets-of-pairs rather than as groups-of-four –
and of those, many had exhibited the peculiar gene-shuffling
phenomenon called 'crossing-over' so that their genetic
effects would not be traceable for generations except by the
laborious process of chromosome-mapping, and even then
only by someone who knew the fundamental secret Dr Fred
had written on this page.

Maurey fingered the sore spot on his ear-lobe, the place
from which Dr Fred took his periodic biopsies, as he took
them from all his 'children'. The spot stung to the salty perspiration on his fingertips, and his whole body was shaking
with fury and frustration.

The tetraploids were not the end of the story.

There was another form to come. Sena was the beginning
of that line – and there was no telling where it might lead, no
telling how thoroughly the children of Dr Fred's tectogenesis
might antiquate the giants. Sena looked like a tetraploid –
but her children would be—

What might Sena's children be, if she were allowed to have them?

The puppy said 'Urrgmph' and hit the floor on one shoulder. She waddled over to Maurey and fell over on her back, requiring that her tummy be scratched; her tiny pink paps offering promise of thousands and thousands of triploid puppies to follow her—

Or, perhaps, tetraploid puppies, with sex-linked double-diploid characteristics hidden within them, to surprise their antiquated tetraploid masters . . .

With a growl Maurey snatched up the projector. The Brobdingnagian puppy, her chunky body all unknowingly the germ of Maurey's plan for the triumph of the giants, and the symbol of his and their defeat, rolled over and crouched, laying back her ears. The force-beam struck the stone floor at her side and pitched her across the room. She got up, barking excitedly, rump high, front paws spread. This time the reflected beam caught her directly under the chin. She screamed and brought up against the far wall.

Maurey laughed and turned the reflector to spread the beam. The puppy regained her courage and charged him, and Maurey broomed her back against the wall again. *Supersede the tetras, eh? We'll see which weapon can be drawn the fastest!* He tumbled the dog this way and that, herded her into the waste-basket, rolled the basket across the floor, overturning it, tumbled the yelping animal scrambling into a corner and out again—

'Maurice!'

Trembling, Maurey let go of the plunger. After a moment his eyes came into focus amid a haze of scalding tears.

It was Dr Fred. Of course; no one else called him Maurice. The geneticist stood in the doorway. The dog whimpered and crawled toward him, her eyes darting back to Maurey in puzzlement.

'Maurice, what – I heard the poor puppy a block away.

What is that thing? Are you trying to kill her? And you've got my safe open! Have you lost your mind?'

Carefully, his fingernails digging into his palms, Maurey said, 'I wasn't hurting her, Dr Fred. It was just a game – she was having as much fun as I was.' He realized that he was holding the silver muzzle of history's deadliest weapon directly in line with Dr Fred's stomach, lowered it with enforced casualness and laughed. That laugh came hard. 'Admittedly she sounds like she's being murdered, she's so damned big—'

Dr Fred strode past him while he was still talking and bent over the stacked papers on the table. 'Why did you open my safe?'

Maurey gave him the prepared theorem. The rangy old man grumbled, almost like the puppy. 'I can see that,' he interrupted. 'Who gave you the combination?'

Nothing would be more suitable here than the truth, Maurey decided. Dr Fred would be interested, probably diverted, by the sharpened talent, and in any event it would be unsafe to tell him that some other person had given out the combination. He might trace the story.

'Really?' Dr Fred said.

He riffled through the papers until he found Maurey's dossier and pawed for the accompanying chromosome charts. 'I wish you'd told me before,' he said petulantly. The charts apparently were mislaid. 'Did you shuffle – no, no, they were an awful hodgepodge before, I know. I really need a secretary, but they're all so bubble-headed. Come see me next Wednesday, will you, Maurice? I want to see if I can trace that auditory acuteness. I *do* wish you'd told me before.'

'I just noticed it myself a little while ago,' Maurey said. His mind was now completely at ease, but his body was trembling; it was an inevitable reaction and it did not bother him.

Still it would be a millennial day when the giants need no longer play-act with Dr Fred—

And it would have to be soon.

The needle, hung by its point, swung back and forth before the window with the regularity of a metronome. As it passed the central pane, Dr Fred stabbed the waxed end of a thread into its eye and jerked it out again. Back and forth. In and out.

After a while he was satisfied that his nerves were all right. He was mad clean through, and he was too old to risk such strong emotions; every hypothalamic disturbance impaired his co-ordination upon which microdissections and chromosome manipulations depended.

He bent to examine the dog, who had lain her chin across his shoe-zipper. She seemed to be all right; Maurice's 'toy', whatever it had been, hadn't hurt her, though certainly it had scared her. Dr Fred wondered what it was. Something electrical, by the look of it. In the old days sadistic kids had shot ammonia into dogs' eyes with water-pistols. Nowadays they hitched the poor beasts to spark coils or something even more elaborate; but in the end it was just the ancient tin-can torture. It didn't even make any difference whether the kids were tetraploid or just ordinary diploid kids; they satisfied their power-fantasies with equivalent cruelties.

He stood up, correcting himself. It *did* make a difference, of course. The giants, even the best of them, had to live in a world which was actively and pointedly hostile; the diploids, except for some few – if much publicized – minorities, had only the general wastefulness of nature against them. Earthquakes hate nobody; but the diploids—

The diploids hated the giants, as well as each other, and had the means to implement it. The psychology of that hatred was obscure; field tests had tended to show that the obvious sources of diploid jealousy – the longevity and the almost incredible physical toughness of the giants – aroused only the most remote, the most intellectualized dislikes: the thalamic disturbances, the hatred that really chewed into the guts, was directed toward the tetra's size first of all, and then toward the makeshift social systems their near-sterility had forced them to contrive. Subconsciously, perhaps, the average diploid wanted to be a giant, and felt himself frustrated; yet – let his children be tetras? Never; no advantage could compensate for the stigma of being so different.

And there were stories of another sort, springing up out of the oppressive sense of sexual inadequacy the giant women aroused, aided in their circulation by the limited fertility of the tetraploid organism. *You know what they say about their women? Free-martins, that's what. A fellow I know told me* ... one of a thousand scabrous jokes. Also, there was a predatory type of woman – not always unmarried – for whom the tetraploid men were natural prey. There were jokes about that, too.

The emotional disturbances among the giants were becoming more and more pronounced as the pressures increased. This tormenting of a harmless puppy was the most upsetting phenomenon yet. It had done more than shock him; it had shaken the very basis of his plan for the giants, as a temblor worries the foundations of an old and solid house.

The swinging needle slowed gradually, its path turning with the Earth's rotation while Dr Fred pondered; at last it hit the window glass and twisted to a stop. The tinny impact reminded him of why he had put it up there. That shock, that moment when he had seen Maurice's beautifully balanced mind wobbling towards paranoia, had frightened him more than he liked to admit. It had been reassuring to find that it had not troubled the neuromuscular co-ordination which was his stock in trade.

But the essential, the ideational shock remained. If the best intelligence among the giants was inclining already towards the easy excuse of persecution, if it had already tipped far enough to fall into the compensation of sadism, then the plan for the tetras which Dr Fred had evolved was too long-term to work.

It was the realization which had reduced him to talking like a senile old man, like a soap-opera doctor, before Maurice, in order to conceal his fear. It had been pure foolishness to pretend that Maurice had disturbed the order of the papers. Nothing went into the safe which had not been culled too thoroughly to require the dim-brained expedients of mechanical filing. He hoped that the giant hadn't noticed; he'd seemed mightily upset himself – a mildly hopeful sign up to a point. Of course a sense of guilt has a threshold; at a certain level of intensity it begins to confirm the habit-pattern rather than inhibit it—

But Maurice had been sorting the papers, too. And Maurice, though he were as mad as the Hatter, was the most alert of all the tetras. He might have seen Sena's papers, and understood them.

He would have understood them had he seen them. And he would have understood, then, some of the ingredients of the time-bomb Dr Fred had planted beneath giants and diploids alike.

Maurice would not be able, now, to await the explosion. He would be perfectly ready to kill Dr Fred to snuff it out—

Except that killing Dr Fred would not snuff it out. There was only one death which would de-fuse that bomb. Dr Fred spread the documents out over the table with a broad sweep of both hands. The code symbols relating to Sena leaped out to his eyes; he snatched the fascicles out of the fan and riffled through them. Chromatin records – molecular film analyses – genealogical summary—

The somatic record was gone.

Whatever might come of actual genetic mutations, the type theorists called cataclysmic, there was implicit in Sena the final flowering of the possibilities of *Homo sapiens*. Those possibilities were all implicit in her somatic record, the first full-length portrait of humanity-to-come. And report and possibilities alike were in the hands of Goliath of the Philistines – a giant and—

A madman.

Dr Fred considered the tears flowing along the creases beside his nose with bitterly academic interest.

IV

Methfessel closed the locker-room door, shot the bolt, and pointed across the low-ceilinged room. The gesture was unnecessary; the golden battle uniforms compelled attention in the drab cement enclosure like a fanfare of clarions. There was one suit hanging in each dull-green locker, tenantless, yet perfect and beautiful with a life of its own.

Maurey strode to the nearest one and examined it with admiration. This basic part of the armour was a heavy breast-plate, hinged to close over chest and back like the carapace and plastron of a turtle. At the bottom of the plate was a brief metal skirt of overlapping leaves, serving as both guard and belt; a control box, mounting a single large red master button and four small black ones, was placed to hang over the left hip of the wearer, and on the right was a holster of plastic straps. The gun in that holster was in some respects like the first projector of the one-way push – but compressed, trimmed, balanced into a proper side-arm.

'No cold-flask?' Maurey said, hefting the gun. 'I see there's an input lead to the control box there; that'll make it a bit awkward to manipulate.'

'You couldn't prove it by me,' Methfessel said, shrugging. 'I sent your figures and the prints to Kelland and he designed this stuff. I wouldn't know a cold-flask from a hot rock.'

Maurey grunted and put the pistol back into the webbing. He was none too sure that he approved of the whole idea of a force-pistol, anyhow; it seemed a trifle overt. Maybe a lance would have been better after all. But the thing was done.

Anyhow the real wonder of the armour was the thing which hung poised on the back-plate, like an eagle spread to abduct a lamb. In some respects it resembled a comic-strip 'flying belt' – which it was – but it was elaborately feathered with delicate flanges, tilted to apply the one-way push over the greatest possible area without at the same time materially increasing the air-resistance, and so seemed to have more 'futuristic' superfluous ornament than anything ever drawn by Dick Calkins.

Maurey was confident that the device did not have a non-functioning square inch, whatever its appearance. Every other flange broadcast the reactionless energy; the alternate flanges picked it up and turned it into mechanical motion. Warping fields, designed to waste controlled portions of the energy as the dove wings of the archaic *Taube* monoplane wasted flying speed, did away with any need for direct methods of steering; they were actuated by movements of the wearer's body. The device had no moving parts, but it would steer with great delicacy.

Finally, there was a helmet, or more exactly, a casque. It was linked electrically with the flight mechanism and with the force-pistol; Maurey could not decide at once what it was for – his sketch had called for a fishbowl. This was topped by a short spike, flanged as if to bleed off heat or some other kind of waste radiation, but that seemed to be purely decorative. It was a true Buck Rogers touch, jarringly out of key with the known superbly functional design of the rest.

'What's that?'

'For protection,' Ira said. 'I don't know why. Maybe you're supposed to ram the opposition. I don't know anything about this kind of thing, Maurey.'

'Kendall didn't say how it "protected"?'

'Not to me. It's called a "transcaster" on the prints.'

There seemed no present answer to that puzzle. It would

take a thorough examination of the circuits to decide why Kendall had put that little Christmas tree on top of the casque. Maurey suspected that the reason would turn out to be as exciting as the rest of the apparatus; the armour was a work of genius; it was a good thing, considering the amount of understanding it evinced, that Kendall was a giant and from Maurey's point of view a dumb giant at that.

'What gets me,' Ira said in an aggrieved tone, 'is that the guy didn't allow any protection for the legs or arms. I'd of thought he'd design a complete body-armour while he was about it – like our football uniforms.'

'He must have had a reason,' Maurey said abstractedly. 'By the way, did Sam Ettinger come to see you yet?'

'That's the bird that walked off the field when Hammy was hurt. Yeah. I took him back in, like you said; but I don't mind telling you I don't trust him. He's a malcontent.'

Maurey smiled crookedly. 'I don't know any well-adjusted tetras, Ira,' he said.

'That's not what I mean. He's what we used to call "disaffected" in the Last Last War. Too full of ideas of his own to follow orders. Oh, well, you're calling it.'

'Either way, you make money,' Maurey pointed out. 'Which reminds me: make me out a cheque for twenty-five thousand.'

'What the hell for?'

'I want you to buy the dormitory grounds where we're living, and that's the price the university set.'

'Think again,' Ira said flatly. 'What do I want with those firetraps? I'll need all the cash I've got to enlarge the stadium.'

'You'll make the money back at the first tournament,' Maurey declared. 'And as long as the university is housing the giants, it's got entirely too much say about what they do. Don't you realize that it'll be sure to prohibit the tournaments as soon as the word leaks through? *Then* you'd be in a nice mess.'

Methfessel thought about it. It was obvious that he was aware of Maurey's additional reasons without knowing what they were; but where money was involved, he was unlikely to ask for a second reason unless the first one seemed insufficient.

'I've got to act as if you knew what you're doing,' he said finally. He pulled out his cheque book. 'Here. Just be sure you spend it all in the same place.'

Dr Fred blew his cork. 'I don't understand you any more, Maurice,' he stormed, pacing back and forth. 'This is far and away the most highhanded thing I ever heard of. The university's relations with that promotor already are quite dubious enough – this modern system of semi-professional handling of college sports is vicious, in my opinion. I should have supposed a research project like ours, at least, to have been immune to such exploitation, yet here you are actually encouraging it – and what's more, turning us all over to a commercial scheme like so many potatoes!'

'Nothing of the sort, Dr Fred,' Maurey said patiently. 'I didn't make the offer, and I didn't accept it; the only choice I had was to refuse to act as the go-between, in which case somebody else would have been found. I agree with you that the gap between the university and Ira Methfessel isn't wide enough by several miles, but I chose simply to be realistic about it.

'As for the tournaments, I take exactly the opposite view (nor were they my idea either, by the way). They give us an opportunity to make a living, and quite a good one, without being dependent upon university charity, and that's something we've all needed a long time for the sake of our self-respect. Granted that it's not a very dignified living, but we're in no position to be so choosy.'

Dr Fred stopped his pacing and looked steadily at the lounging giant over the tops of his glasses. 'You haven't said anything yet that I don't think a half-truth or a plain fantasy,' he said. 'I'll pass your curious conception of "realistic" behaviour; that kind of expediency is no novelty to the world, heaven knows. For the rest – evidently you think that by making the tetraploid project over into a business venture you weaken Methfessel's direct hold on university affairs. That's a half-truth; the other side of the coin is that at the same time you've ruined the university's reputation more thoroughly than could any possible relationship with professional sports. Twenty-five thousand dollars! What can the Board have been thinking of? They might just as well have sold the whole chemistry department, students included, to Columbian Pharmaceuticals across the river!

'No, don't interrupt, Maurice. That's only the beginning. I suppose you realize also that this terminates the tetraploid research as far as I can be concerned. Now that the giants are living on privately-owned property, they will have to be treated like any other graduate assistant or scholarship student. The social aspects of the study go out the window, nor will I have any authority to direct or in any other way interfere with the genetic course of the experiment. I'm left with nothing but a group of volunteers.'

He snorted. 'Volunteers! No wonder the so-called "social sciences" were a bust! "Will one or two atoms of oxygen kindly step forward and answer a few carefully impersonal questions?" '

Maurey started to speak, then halted before his lips had parted. Best to let the old man talk it all out.

'What the reaction of the public will be to this I don't know, but it's bound to be bad,' Dr Fred went on, a little more quietly. 'I can only hope that it won't be extreme. You've expelled your brothers into the status *private citizen*, and you're going to find that that's a much more dangerous and humiliating status than the one you had before, the one the public equated with that of experimental animals. No, your motives may be good, Maurice – though I shan't say I believe it – but good or bad your actions have been despicable.'

Dr Fred turned his back on Maurey abruptly and stared out the smudged window.

'I rather expected some such reaction,' Maurey said evenly, 'but I certainly didn't anticipate so much violence. It's unlike you to use such loaded terms, Dr Fred. The mere fact of a change of ownership of some property is not going to change your relationship to us in the least. The tetras all owe their existence and their biological advantages to you, and they won't forget it. They'll still be exactly where they were before; the fact that that bit of land can't be called "campus" any more changes nothing on the event level.

'They'll still come to you and co-operate with you in your experiments. They'll look to you for guidance as before. Your anathemas on "volunteers" do nothing but hide the fact that we were always volunteers; you never did have any dictatorship over our personal lives, nor did you ever try to exercise any – *not until now* – so it seems foolish in you to complain that any real control has been taken away from you. What you had before, you had now.'

'Yes, yes, Maurice,' Dr Fred said wearily. He did not turn. 'I am not as unaware of the difference between real meaning and formal meaning as you'd have me think. Your speech tells me, for instance, what you see: a feeble old man querulous over the intervention of cold reality between himself and his pet hobby. I withdraw my implied accusation of bad faith, which I had no right to state; but I believe you are as aware as I am that what you have done will have evil consequences, for the giants, Maurice – not for me, but for the giants. Incidentally, what did you want Sena's somatic record for?'

'I didn't want it. Am I supposed to have it? And what has it to do with what we're talking about, anyhow?'

'I would very much like to know,' Dr Fred said. 'However your course plainly does not include me, so I shan't pursue the matter. You may as well go, Maurice. You can't undo what you've done in any case, nor can I expect to talk you out of whatever it is you're planning. But I will tell you this: Your sanity is dubious.'

'I'll go,' Maurey said. 'Since we've reached the wild accusation stage already, the matter is closed. Good-bye, Dr Fred.'

He closed the door with precision and went down the squeaky old wooden stairs. On the whole, though he was a little angry – that couldn't be helped – he was well satisfied with the way the interview had gone. Most of the giants, he was sure, would be glad to be out from under the fatherly eye of the university; they would come to Dr Fred, of course, and find the old man accusing Maurey of some heinous plot the details of which he would not be able to give.

The end-product would be a strengthening of his own already considerable influence (since, although he would deny engineering the transfer, the impression of himself as prime mover would remain) and an abrupt slump in Dr Fred's prestige. Of course, Dr Fred might not react as predicted, but in view of his present performance that danger was small, and in a few days it would no longer matter what he said or did.

He stood for a moment on the stone porch of the building, looking over the campus. It was late in the afternoon; probably Sena and Sam would eat at the student cafeteria. For a moment he was prompted to let the matter wait until tomorrow, but then thought better of it – providing properly for the future entailed so many unpleasant tasks that were each one postponed only a single day the future would never arrive at all.

He drove the roader over to Sena's dorm, left a note for her, and then proceeded to the men's dormitory, where he let himself into Sam's room with a passkey. Sam's typewriter had a half-finished letter in it, but its content was uninteresting. Maurey selected a book and made himself comfortable.

Sena arrived first. She was somewhat flushed. 'I got a terrible ragging from some of the kids on the first floor,' she said. 'Are you sure it's all right for me to be here, Maurey?'

'Quite sure, my dear. I'll explain as soon as Sam arrives. I must warn you that the situation's rather complex – and not without its unpleasant aspects.'

'Really? How ominous!' Sena sat down on the edge of Sam's cot, raising her eyebrows in mild alarm. 'Can't you – oh, that sounds like Sam now.'

Maurey's supernormal hearing had already detected Sam humming under his breath at the foot of the stairwell. When he entered, the black-haired giant's surprise was comical. 'What are you trying to do, you two, get me thrown out?' he demanded, half seriously. 'After I went back to Maurey's crazy football team? There's gratitude for you!'

Maurey grinned and explained what had happened.

'So this is a private apartment now, Sam, and you can have a woman in it if you like.'

'I like,' Sam said immediately. Sena smiled.

'I thought you would. But there's a hitch. I knew that as soon as you heard the news, you and Sena would want to follow through on your commitment – your housing problem is already practically a legend; so I wanted to talk to you both, and try and persuade you not to do it.'

'*Not to* do it?' they said together. Sena leaned forward. 'What do you mean, Maurey?'

'Sena, how much do you know about yourself – about your genetic make-up, that is?'

'Why, not a great deal,' she admitted, frowning. 'About what we all know about ourselves. I know who my parents were, and that one of them was related to Dr Fred, and I know the theory of chromosomes-doubling.'

'That's what I thought. Do you know any more, Sam?'

'About myself?'

'No,' Maurey said. 'Sena.'

Sam shook his head, patently mystified. Maurey paused a moment. He realized that he liked Sam, and he wondered if it were really necessary to be so brutal. The two kids loved each other; wouldn't it be enough to persuade them not to have a child?

He realized at once that the suggestion would be just as badly received as his earlier one. The near-sterility of the giants had made birth-control close to a crime among them; and besides, what if there should be an accident? A tremor of pure terror made him catch his breath.

'I hate to say this, but it's got to be said,' he declared. 'Sena, I've seen your records; Dr Fred showed them to me. And you're *not a tetraploid*.'

Sena went white, and one hand flew to her throat. 'I'm – not?' she said faintly.

'I'm afraid not. Essentially – forgive me, both of you, but this thing transcends all of us – essentially you're a diploid. Your size is tectogenetic in origin. You were given that one characteristic by direct manipulation – one of Dr Fred's famous micro-operations on the genes. If you and Sam have a child, it will be triploid. Like Decibelle.'

'Are you sure, Maurey?' Sam said slowly. 'Why would Dr Fred pull a trick like that? He told me that the dog was strictly an experiment, and that he hadn't gotten around to testing triploidy in humans.'

'And so he hasn't. But when Sena's child is born, there's his test. As for why – well, naturally scientific curiosity must have been one reason. He had to provide a diploid human being to mate with a tetraploid, so naturally he had to supply one of practicable size.'

There was a brief silence. At last Sam said, 'Maurey, I'm sorry, but I can't swallow that. Dr Fred's not underhanded; he would have told Sena, and when he knew who Sena was going to live with, he would have told him, too. In this case, me.'

'That's right,' Sena agreed. 'You must have misunderstood him, Maurey. After all you're not a geneticist, even though you are our biggest brain.'

Maurey shook his head. 'There was no reason for him to tell anybody. You've seen for yourself what a Blue Ox that dog is; would you have known it for a triploid if Dr Fred hadn't told you? Of course you wouldn't; and your child would look like a tetra, too, just as Sena looks like one.'

'But the reason, Maurey, the reason!'

'I can't be sure,' Maurey said. 'But Dr Fred's an old man, and he doesn't think as straight as he used to. When I objected to this whole business he turned on me in a white rage – I was flabbergasted, let me tell you. The first thing I wanted to know was why he couldn't have implanted tetraploid germ cells in a diploid woman, artificially, instead of creating all this incipient heartbreak. That was when he lost his temper, so I never did get any answer.

'However, it's my guess that he doesn't think the tetras to have been a successful experiment, and so he's planting ringers among us. Sena is almost surely not the only one. In a few generations we'll be cut back down to diploid size again – we'll *be* diploids – without ever knowing exactly how or why it happened. And nobody will try the experiment again – because by that time the existing laws against further chromosome-doubling in human beings will be given a full set of teeth. Hell, I'm not even sure *I'm* not a phony – you can imagine what a shock this was to me. I think I can claim to know just how you both feel. But, there it is.'

Sam swore and sat down abruptly, feeling for the arm of his desk chair with the gesture of a man gone suddenly weak in the knees. Sena was blinking, unsuccessfully trying to squeeze back tears. Maurey felt simultaneously like a house and like a composer whose sonata has just been afforded an ovation.

'Of course it's hard to take, damned hard. I don't ask you

to bolt it down whole, or expect you to; as Sena says, I might have misinterpreted what I saw; that might just possibly account for Dr Fred's being so angry with me. I'd be delighted to be proven wrong, believe me. One of you ought to check me.'

'I'll talk to him,' Sam said. 'I can't quite see you telling us this if it isn't so, Maurey, but of course we'll have to be sure you haven't gone off half-cocked.' His voice wavered dubiously as he reached the end of the sentence. 'Damnation! We'd really be out in the cold if it's true.'

'Life is aye full of cark and cauld,' Sena said. The attempt to be cheerful was pitifully futile. 'I won't get *any* man, big or little, if I am just a phony. But Maurey, I'm nearly forty years old, and I was just getting out of an awful ten-year adolescence when I was twenty-eight. Doesn't that disprove your theory?'

'I wish it did,' Maurey said sombrely. 'But unfortunately it doesn't prove anything either way. You'd have to be long-lived in order to seem like a tetra, so Dr Fred might have given you that characteristic too, for the purposes of the masquerade. Or you might be a true tetra, and I've caused all this fuss for nothing. I can only say that I wouldn't have opened my mouth if I hadn't been close to positive about it. All I want to know now is this: do you agree with me that, if it *is* true, Sena should not – should not bear any children?'

'No,' Sam said. His voice was gravelly. 'For all we know, Dr Fred might be right: the tetras might have turned out to be an unsuccessful experiment. I'm convinced that you mean well, Maurey, but I won't commit us to anything until I've checked.'

'Fair enough,' Maurey agreed, rising. He was glad that Sam had chosen to be stubborn; it banished the last traces of that momentary regret. Sam was thoroughly likeable, but in this chess-game no piece was indispensable.

'Check,' Maurey said.

V

By the next day the story somehow had become common stock among the giants, though Sam was reluctant to believe that Maurey had been circulating it in advance of conclusive evidence. Possibly some part, if not all, of the discussion in the dorm had been overheard – voices had a way of leaking out under doors, and the cement stairwell made a passable whispering gallery; nor could any tetra be condemned for eavesdropping upon a matter of such intimate and personal importance to all of them.

At first the reactions varied widely. There was flat incredulity—

'Sam, who's mad at Sena? Somebody's spreading the gawdamndest fairy tale!'

and reluctant sympathy—

'Tough luck, Sam – it must be pretty rough on you, too.'

as well as immediate chauvinism—

'Good thing you found out in time, eh, old man?'

There was also a startling amount of covert hostility: some of the giants went out of their way to avoid the embarrassment, or the humiliation, of speaking to Sam.

It was still worse to find that a few of the giants had not shucked off the predatory vices of the diploids with their size. Several of them, Sena reported, seemed to have concluded that Sena needed 'Consolation' and would therefore be easy pickings. This shameful reminder of a common, ignoble ancestry troubled Sam most, although he could not say quite why he found it so ominous.

And nothing could be done about it. Dr Fred was out of town, attending a world congress of geneticists in Toronto. As the month went by, Sena's presumptive diploidy receded gradually as *the* subject of conversation among the giants, and was replaced by a gathering excitement over the new private-citizen status and the Paying Job. There was also a good deal of speculation over a possible revival of the tournament idea, though neither Methfessel nor Maurey had mentioned it in over a month.

Underneath all this Sam saw the reactions to his and Sena's problem begin to divide and flow away from one another in two definite streams. The unbelievers and the sympathizers showed a tendency to merge into a common camp of support for the outcasts; while the chauvinists, the suspicious, and the rejected wolves clumped together elsewhere, more slowly, like bloodcells in an antagonistic serum.

Even grimmer portents were visible to Sam, whose deep personal involvement had sensitized him to the slightest signs of new trends. The division among the giants began to express itself in terms of the two teams on which they were now earning their living.

Tetras sympathetic to Sam and Sena, naturally enough, predominated in the Titans, where Sam played. As a result, the members of the minority faction began drifting over to the Atlanteans – where the same phenomenon was taking place in reverse. Methfessel, who now managed both teams, did not attempt to block the exchanges; indeed, Sam suspected him of encouraging it.

Certainly it was to Methfessel's advantage, for it brought the rivalry between the two teams – heretofore only a desultory, token rivalry at best – to a state of real acrimony, and the games became rough almost to the point of viciousness. The crowds loved it. The games always had partaken of the spirit of a mass gladiatorial contest – a spirit which is entirely a function of the temper of the spectators, not of what specific game is being played on the field – and now the players were accommodating themselves to the mood. The gate increased at once; the stands were packed for almost every game.

And the percentage of on-the-field injuries increased enormously.

Before Dr Fred came back, it was already too late to

scotch the schisming of these two camps even with any statement that there was nothing to Maurey's soft denunciation.
Something had happened – Sam could not find out what –
which had blighted Dr Fred's authority among the tetras;
they spoke of him in a way they had never spoken before, in
a tone which regardless of the words had contempt beneath
it. Sam tried all day to reach him, but Sam's own rigid
schedule was in the way; he reached the scientist at last, by
phone at four in the afternoon just before a field trip, and
then the old man evaded Sam's necessarily cautious questions and asked to see him at once, which was impossible.
Sam had to settle for an appointment at 6.00 a.m. of the next
day.

He thought of spending part of the penultimate evening
with Sena, but knew at once that it would be the worst move
he could make. Their situation was already dismal enough,
without their sitting for two or three hours staring miserably
at each other trying to find something to say, or proffering
each other comfort where there was as yet no rational re-
assurance to be had.

He called her, and told her of the appointment. Her quiet
understanding made him feel a little better, but a moment
later he was doubly aware of how desperate he had become.
He was already grasping at the smallest straw.

Being alone in his room was even worse. He could not
concentrate upon his technical books for more than three
minutes without becoming conscious all over again of the
all-gone feeling in the pit of his stomach, and the troubles of
fictional characters filled him with a furious impatience:
Emma Bovary had enthralled him for years, but now she
seemed like a fool who had invented troubles in the absence
of any real ones. At midnight he had enough and threw
himself out of his room without stopping to lock the door or
put out the goose-neck desk-lamp.

The long, aimless walk through the dim campus brought
him finally to the edge of the river. He sat down on the

steep-sloping bank and began to chuck stones into the black water. Each stone distorted the reflections of the lights of Columbian Pharmaceuticals on the other side, turning them into cold wriggling flames. After a great while he stopped throwing rocks and just sat, hugging his knees. The circling of his own thoughts numbed him, and the images on the water writhed hypnotically without any help . . .

Across the water there was a shrill, mournful hooting. He blinked and sat himself up straight, feeling cramped and emotionally washed out. The hoot, he realized slowly, was the plant whistle, calling in the third of the firm's staggered shifts. That made it 4.00 a.m. His watch confirmed it.

Might as well walk slowly over to the Genetics Building and wait for Dr Fred to arrive. The wait would be tedious now that he was more awake, but some time could be killed by cutting through town and picking up breakfast at an all-night beanery. Unfortunately, he was not hungry. He climbed the sandy bank and began to walk, favouring his stiff muscles.

It was already dawn by the time he came in sight of the building. No one was stirring. It seemed a shame that such a peace ever should be broken, spurious though it was. He went up the broad river of steps, paused, and went inside, where it was warmer; he was chilled through.

The door of Dr Fred's lab was ajar. Before Sam touched it he could see that the safe was standing open. Papers were tumbled out of it in a frozen cascade. His stomach-muscles knotted.

A robbery? But what did Dr Fred have that anyone would want to steal?

He felt the answer searing its way up towards the surface of his mind. Anything was better than having to face it. He lunged through the door.

His first impulse after that was to run headlong back the

way he had come and throw himself into the dawn-bloody river.

Dr Fred was tumbled grotesquely on the boards, half under the workbench. His cheek and shoulder rested in a sticky black pool. In spite of his twisted position, it was easy to see that his entire rib-cage had been smashed in by some single, unimaginable blow.

Decibelle growled; then, recognizing Sam, she lifted her chin from the dead man's shoe. Whining softly, she began to crawl towards him on her belly. Sam bent abstractly and put out a trembling hand towards the dog, but his eyes had already found the weapon and could not leave it.

It lay shattered in the farthest corner of the room, the one that was always darkest during the work-day. Now it was directly in the merciless early sunlight; and, despite its almost total breakage, he recognized it.

He had made it.

It was the projector of the one-way push.

All but a small percentage of Americans live out their lives without ever coming closer to murder than the daily tabloid can bring them, though magazine fiction and video confer a spurious intimacy with the subject. Sam was no exception. To say that he was overwhelmed with horror and fear is to say nothing, for, although true, the phrases did not correspond with the feeling: the emotions he suffered were horror and fear but they were entirely unlike any emotions he had ever before associated with those words.

He realized that he should be doing something, but nothing occurred to him that was not wildly irrelevant. He simply squatted, absurdly scratching the half-grown dog and trying to think – not a rational thought, but just any thought at all. His whole mind was fragmented. Perhaps the most terrifying thing still was that instant searing flash he had felt at the moment he had first seen the body; that stab of *guilt*. Traces of it still remained, unexpungeable by mere certainty that he was innocent.

Partly, of course, the guilty feeling had come from an underlying consciousness of being in a bad position. Dr Fred had been murdered while Sam was sitting by the river, alone, unable to account for his time; he had left his room light on, which would look like an amateur's attempt to establish an alibi; and a motive could be shown, a motive stronger than many a one which had hanged accused men before.

But the sensation had been stronger than simple fear. It had had all the flavour of conviction, of a compulsive self-knowledge: '*I* did it.' It had brought out the buried guilt of the outgroup, of the man on the defensive, the man whose real guilt is that of being different from his fellows.

All these fragments fluttered confusedly inside his skull for over two seconds. His first formulated thought was: Would a statistical study of the neurotics who run to the cops with 'confessions' of every publicized crime show a predominance of minority-group members?

The question was so remote from any 'proper' reaction to murder – as such reactions were taught in the video school – that he could scarcely smother an hysterical giggle. But it freed him. He found that he could think again, with at least passable coherence. He gave the huge puppy a final pat and stood up.

It would be at best futile, at worst damning, to sneak out and let someone else discover the pitiful corpse. He was fairly saddled with it, and the real killer had planned nothing else; this much had to be accepted at the beginning. Sam knew that he could not hope to outplan such a man from a standing start. He would have to consolidate his position within the frame of a probable death-cell.

He had one advantage. The killer could not have antici-pated that Sam would find the murder out at dawn, unless he had tapped Dr Fred's wire and had so learned of the early-morning appointment. That would require vigilance of an order which Sam was convinced was impossible for anyone who needed to pretend to a 'normal' schedule at the same time. Probably it had not even been planned that Sam himself should discover the body – accident had given Sam nearly the worst possible set of circumstances, but accidents cannot be planned; the crime had been expected to speak for itself, in Sam's absence. The killer, in short, could hardly have expected that Sam would be able to investigate before anyone else.

It would be half an hour at least, Sam estimated, before the first assistant professor or instructor would enter the building, and at least an hour before the first undergraduate would be seeking Dr Fred's lab and advice. Fifteen minutes should be enough to examine the sludge of papers before the open safe, and the opportunity justified almost any risk.

Sam pulled the sleeve of his jacket down over his left hand and slid open the draw where the gynaecological equipment was kept. The rubber gloves were there, all right, but they were dulled by a thin film of dusting starch. Anything he touched with those would be marked. Yet he could not afford a fingerprint; he had never before touched Dr Fred's records, and it would be important to leave no evidence that he had.

Again he decided in favour of the lesser risk. Use of the gloves would show, but it would not point, except to suggest that there might be fingerprints inside the tips of the gloves. He wished the forthcoming fingerprint experts joy of that problem, for without stopping to think he could name fourteen people who had worn those gloves within the past month alone.

He used the gloves and put them back in the drawer. Sena's entire dossier was missing; so were those of Kelland, Hammy, Maurey, and Sam himself. In addition, random sections of other dossiers were, in his own fervent cliché, conspicuous by their absence. The names of the giants involved made a group in Sam's mind, but he could not quite label the group as yet, and he gave over trying for the moment. The absence of papers on Sena's and his own case was conclusive enough for his own purposes, since it enabled him to name himself the name he had been crowding out of his consciousness up until now: Maurice St George.

Maurey, the chief god of all the stumbling Olympians Dr Fred had produced, had rewarded his maker.

Sam could appreciate the subtlety of the planting even better after that conclusion. The apparent crudity of the frame-up – for instance, the abandonment of the unique, easily ticketable weapon – would seem to rule out Maurey at once. Maurey had done more than implicate Sam: he had staged the scenes to suggest a clumsy attempt to *fake* a frame-up. Sam felt an iron-cold certainty that Maurey's efforts would not go to waste. Maurey never did anything incompletely.

Was there anything more? Yes – the dog. There was one remaining factor visible to Sam upon which no plan could count. Maurey could not have dared to kill the dog, since it was known among the giants that Decibelle did not like Maurey; furthermore, Maurey was unsentimental, and would not have thought of the dog except in terms of telltale torn trousers. Instead, Sam surmised, he had worked quickly, well above the level of the animal's understanding until it was far too late; and had then left, before Decibelle had confirmed Dr Fred's death or the fact of a quarrel. Very probably there had been no quarrel, but only an unexpected silent blast from the projector, a crash of equipment, a heavy impact, footsteps receding in the creaking stairwell – and a frightened, an abruptly and puzzledly lonely dog.

But the dog was not stupid. It was not in any sense an ordinary dog. Maurey had gotten away without being attacked, but there were now some matters about which no doubt could exist in Decibelle's slow but inexorably direct mind.

'Decibelle,' Sam murmured. 'Hey-pup. Where's Sena? Where is she? Where's Sena?'

Decibelle looked up.

'Hey-pup. Hey Decibelle. Where's Sena? Find Sena.'

The repetition told. The dog, her ears still drooping, looked towards the door, and then back at Sam. 'Go find her. Go on. Go find Sena.'

Decibelle thought about it, blinking her bloodshot eyes alternately and rather upsettingly. Then she stood up, or almost stood up, and crept back toward Dr Fred.

'No, no. I'm here, puppy. I'll take care of Dr Fred. Don't you worry. Leave it to Sam. Get Sena. Come-on, Decibelle, find Sena. *Hey*-now. That's a good pup. Go to Sena. Go find Sena.'

The immense animal looked back at Sam.

'That's it, Decibelle. Sam's here. Go on Sena. *Come*-on, Decibelle. Find Sena. Tell Sena about Maurey. You're a good girl, you did your job. Now tell Sena.'

At the word 'Maurey', the hair along the dog's spine coarsened. By the time Sam had come to his final order, Decibelle's back looked like a scrubbing-brush. Her claws ticking on the boards, she moved reluctantly towards the door.

'Hurry, Decibelle. That's it, that's it. Find Sena! Quick pup! Go get Sena!'

Suddenly, it took. The dog growled, softly, a sound as ragged and ugly for all its distance as the encounter of a buzzsaw with rusty metal. Instantly and without transition the great beast bayed, bayed enormous and bloody murder, and lunged out down the stairs. The belling cry burst forth on to the campus and receded on the fresh morning air.

Sam listened to the dimming clamour for a moment. Then he swabbed his forehead with his wilted handkerchief and picked up Dr Fred's phone.

'Get me the police.'

*   *   *

VI

The Civil Freedoms Association met in the cellar of the Romance Languages house, in a moderately luxurious, cedar-panelled room which, though quite small, was usually far too large for the group. Tonight, however, the attendance had turned out to be so large that the cellar clubroom was an impossibility. As the room became more jammed and the air still bluer with smoke, June looked more and more worried, and Maurey, despite himself, more and more contented. Finally the meeting was adjourned upstairs to the building's largest classroom.

The reason, to nobody's surprise, was a turnout in force of the tetras themselves – the diploid membership of the equal-rights block had never been more than tiny. The crisis over Sam's killing of Dr Fred had made a general conference unavoidable; and Maurey had agreed, reluctantly, with June that the giants' strongest diploid supporters should add their small encouragement. There were also a few policemen, supposedly there to protect so large a gathering of giants and fellow travellers from being mobbed by a possibly outraged citizenry; their number was small, but their publicity value was enormous. (This suggestion, too, June had thought her own.) The presence of the cops, in turn, made newspaper reporters inevitable.

The formal opening of the meeting was considerably delayed while Maurey waited for the last possible giant to appear. The rest of the tetras – mostly of the Titan faction – twisted in their seats, like high school students crowded under fourth-graders' desks by a building shortage, and muttered to each other. The UPI reporter, who had also to attend an anti-vivisectionists' conclave across town, interviewed Maurey briefly without seeming to listen very closely to his answers, and left.

Finally June caught Maurey's eye. He shrugged and

moved his fingers away from his chest as if trundling some round object towards the edge of a table. June made a smart tattoo with the gavel.

'Friends, let's get to the business at hand,' she said clearly above the thrum of talk. She looked extraordinarily young on the rostrum. 'I won't call roll or fuss around with parliamentarianisms tonight – this meeting is too important and we're starting rather late. I'm going to ask our large confrères to sit quiet a little while longer while we hear from Tom Drobinski. Tom's editor of the Dunhill *Campus Echo,* the head of our public relations committee; I think he can tell us something about what the public temper is like right now. Shoot, Tom.'

Drobinski, a swarthy sophomore journalism major with a cranial structure that would have thrown a frog into convulsions of jealousy, stood up and said rapidly: 'You've all seen most of the papers so I won't go into detail on that. Briefly, they're all taking the same line, except the *Worker*, which hasn't taken any notice at all yet, and the *Times*, which made a fairly successful try at being impartial.

'We haven't any facilities for monitoring, but the videocasts I've seen myself all played up the parricide angle, and used lots of myth-faking and crude dream symbols – heavy emphasis on mystery, hazy gigantic figures, Biblical references to "giants in the earth", the kind of thing that makes people feel alarmed without knowing why. On the whole I think everybody, but everybody, thinks Ettinger is guilty, but sort of wish he weren't – some kind of identification-reaction there but I'm not analyst enough to make it very clear.'

'Give us a sample, Tommy,' one of the diploids said.

'Well, Bax Ferner has a long, quasi-Freudian chew in tonight's *Weathervane*, hinting that big people are just naturally murderous because nobody loves them, but that it's black Fascism to single them out. But that's on the side. I saw a mess of news service dispatches from the capitol just before I came here. One of the state senators is going to put a resolution on the floor tomorrow to have the tetra colony taken under state supervision now that the university's sold its jurisdiction—'

'But Tommy, that must be illegal!'

'No it isn't,' Drobinski said. 'The laws relating to Indian reservations haven't been needed for seventy years, but they're still on the books. That's only the beginning: there's another resolution being drafted to register all the tetras with the Habitual Offenders Monitor, give them numbers, make them show wallet cards to new employers, and all that. And there's going to be a lot more trouble tomorrow – Ira Methfessel has just announced a big tournament of some kind, evidently the one that all the rumours have been about, and the stadium box-office claims that people are already climbing all over each other to get tickets.'

He stopped speaking as if he had been turned off, and sat down. Then his voice shot forth again from among the seats, startlingly. 'We're about two days away from Pasadena, I'd say,' he declared with flat clarity. 'Only this time—'

'June, may I chip in?' The deep gentle voice came into the tense silence like a benison. June smiled.

'We have to think about Sam right now,' the speaker said from the back of the room. 'Maurey, do you think he has a prayer of getting a fair trial?'

'Yes and no,' Maurey said, rising. 'Obviously there'll be political bias; it'll be impossible to pick a jury that won't already be largely anti-tetra, emotionally. Is that what you mean, Kelland?'

'Just that.'

'Well, I see nothing that we can do about it. Except for that factor, I expect the trial to be scrupulously fair. Naturally, we'll have to get ourselves a good lawyer, as brilliant a man as our pooled resources can afford. I'm sure

Methfessel will let us have an advance on the gate for the tournament if he's properly approached—'

There was a racket on the floor. The centre of the disturbance was another giant, an Atlantean, who was now standing and shouting. Since there were four other giants shouting at him, Maurey, and each other at the same time, nothing coherent came through. June added the crack of the gavel to the din, which subsided promptly.

'What were you hollering, Briggs?'

'That it's ridiculous to talk about doing anything for Sam Ettinger,' Briggs said hotly. 'What he's done makes him as vicious an enemy of ours as the kept press. If we band together behind him the public will identify us with him. What we should do is draft a resolution condemning the murder, and demanding quick, merciless justice; pass it unanimously, and give it to the reporters.'

'That's the stuff to give the troops,' an excitable diploid crowed.

'I have nothing against mercy myself,' Maurey said mildly. 'And neither one resolution nor twenty is going to speed up justice any faster than the law will let it go.'

'The words don't matter; the important thing is to disassociate ourselves from Ettinger.'

'Throw out the Jonah,' Kelland suggested. Briggs failed to take the remark as a criticism, or even to place the metaphor.

'Exactly; throw him to the wolves,' he growled, deporting the seafaring Jonah to a droshky with a single stroke. 'He's earned it. What he's done is untetraploid. He's a – Pasadentist.'

'What was done was bad,' Maurey said, unruffled. 'But we have absolutely no proof that Sam did it. He designed the weapon that was used, to the best of my knowledge, but I myself put the weapon into Dr Fred's safe, and there's no direct evidence that Sam got it out again, or that he even knew the combination.'

'I'm not sure he designed that gadget,' Kelland interrupted. 'I've worked on things of that type myself, from knowledge you gave me, Maurey.'

There was a flurry of scribbling among the reporters. Maurey frowned warningly, but Kelland plunged on. 'You have my drawings and could have built a projector yourself to my design. Hell, Maurey, nobody even knows whether or not the gadget really was the murder weapon. A young male giant would easily crush a frail old man's chest in an identical fashion with the back of a shovel, in one swipe. The projector could have been a blind.'

Maurey could feel himself helplessly going white-lipped. Since he could not control the reaction, he would have to account for it before some dangerous construction was put upon it. The easiest and quickest way was to take offence.

'Pardon me, Kelland,' he said, 'but it's a good thing I know you well and know you're a blunt and sometimes blundering sort of guy socially. Otherwise I might have to lose my temper. Everything you say is true, but it could also be taken to add up to an accusation – of murder. Not a very wise thing to do in public.'

'I'm sorry,' Kelland said at once. 'I had no intention of accusing you. I simply wanted to point out that Briggs is hanging Sam well in advance of any proof that he's done anything wrong.'

'The point,' Maurey said, 'is well taken, if badly put. What about that, Briggs?'

Briggs' opinion was succinct but unprintable. 'I demand a vote,' he added.

'On what?'

'On whether or not we denounce Ettinger. What else?'

'Will you go along with the decision if it goes against you?' Maurey asked curiously.

'Sure; what do you take me for? *Whatever* we do, it ought to be unanimous. You're asking so many questions, let me ask you one: who do *you* think murdered Dr Fred?'

'It's not my function to decide that,' Maurey said, making each word tell. 'However, Briggs, I seriously doubt that any tetra would have raised a hand against the old man, whatever the fancied or real provocation. If we do decide to help defend Sam, part of our effort ought to go also towards looking elsewhere.'

There had been a murmuring of side-chatter all through the meeting, a murmur of private debates whose participants could afford more than one ear to the discussion. Now, as the iron in Maurey's speech began to bite deeper, the room became more and more quiet, until at last there was an unearthly silence. The reporters bent intently over their notebooks, and Maurey could see the headlines being conceived:

LUMMOXES HINT NON-GIANT SLEW DR HYATT: THREATEN VENDETTA

But he wanted those headlines to surprise the tetras, so he did not dare let the silence persist long enough for full comprehension to set in. 'June,' he said, 'would you tear up some paper and pass out the fragments? If you favour hiring a lawyer for Sam, friends, write "*Yes*". If the "*No*" slips predominate, we'll entertain a new motion.'

But, of course, the yesses won. There were two '*No*' votes. One was Maurey's own, cast to break up Briggs' suicidal ideal of unanimity; Briggs himself had agreed to cast the other one, and Maurey was more than glad of the foreplanning. He had not anticipated such a landslide.

He announced the results. The reporters broke for the door. Maurey looked at Briggs, who shrugged. The shrug was a genuine artistic stroke; Briggs would become a great actor in the new world, Maurey thought, if he lived to see it.

Maurey was in honest doubt as to whether or not he should.

Sam had found with dull astonishment what every newly-imprisoned man finds: that after only a brief isolation from his own world, he could no longer understand the news. He read through the local paper's lead-story account of the 'conference of war' with the conviction that none of the tetras whose names were attached to the quotations could have said anything like that; yet, as a whole, the story hung together.

There was a great deal more that was puzzling. Methfessel had announced his tournament – there was a half-page ad for it in the sports section, and half the editorial copy in the section was devoted to it. Methfessel's ad made very little sense:

'SEE Titans in Deadly Combat! SEE Flying Shock Troops Clash in Mid-Air! SEE Affairs of Honour *FOUGHT TO THE FINISH* with SWORDS OF FIRE! Towering Heroes contend for the favours of Georgeous Giantesses with strange Weapons *never before used on any battlefield*! Champions in armour – mass charges – futuristic warfare – blazing colour, beauty, spectacle! THE EVENT OF A LIFETIME!' And more, proving nothing that Sam could see but that Barnum was not dead after all.

The sportswriters were generally hostile, or, at least, sarcastic, but appeared to have little better idea of what Methfessel actually planned than Sam could deduce. Certainly the propaganda hardly suggested mailed knights on brewery horses, despite the medieval trappings of the ad-writer's copy. On the editorial page, the newspaper's proprietors took a dim view of the whole business, suggesting darkly that there was something frivolous – in another century the editorial might have said 'worldly' – in the giants' staging a circus when their whole existence was a matter of the gravest concern among right-thinking normal human beings. Like most editorialists, however, the writer seemed to fear standing too strongly on one side of the fence, despite not having to sign his name; for the editorial wound up with a foggily hopeful remark about people putting their best feet forward. Perhaps this was intended to pass for impartiality.

The smaller stories about the crime itself were a little more comprehensible. Sena, who had been held as a material witness, had been released upon a stiff bail which Maurey had put up. (Sam had been unable to speak to her.) The paper had a 'human interest' interview with her, which wavered nervously between straight sob-sister treatment and a tendency to acidulousness. There was no mention of the dog, for which Sam was grateful no matter what it meant.

The trial date was already set, a two-column italic head announced. A box on page twelve contained an irrelevant datum about Sam's university post, evidently supposed to be funny. There was a column of fuzzily learned speculation about the weapon, written by the man who usually did the paper's 'little Walks With Nature' – Maurey had declined to explain the mechanism except before the grand jury, on the grounds that there was a patent pending on the principle which publication of details would prejudice. Finally, there was a meandering round-up yarn dealing with official reactions to the murder, including the news of Sam's dismissal from the graduate faculty, and a comment from the state's governor which promised prompt punishment for 'provocative acts'.

All of which was alarming without being in the least enlightening. What most irritated Sam – his strongest emotion now, for it was impossible to sustain an intense consciousness of personal danger continuously for a week – was his being cast, at this point, for the role of the Impotent Husband in a bad videocast. The stage was all set for the Big Think, wherein the male lead was to take a long walk or be shut up in a room until he Came-to-Realize, with his own voice squawking at him through a filter (representing Thoughts) to a background of treacly organ-music.

The irritation, of course, sprang from the fact that a Big Think was by this time a commodity for which Sam had no use. He had Come-to-Realize a week ago, in a split second, without benefit of *vox humana tremolo*. Sam's thinking was often slow, but his conclusions none the less sound for being belated. He knew, he was certain, the name of Dr Fred's murderer, and he knew in general what Maurey's purpose was: to widen the gap between giants and diploids by every subtle means, and to provoke an eventual break in which Pasadena would happen again *in the opposite direction*.

Sam had been blind to the implications of the one-way push as a weapon, but his was a type of mind that saw things at once upon demonstration. Jolted into thinking of the phenomenon in military terms, half a dozen expedients occurred to him – side-arms, pressor fields, anti-missile field – any one of which he could have designed with a minimum of experimentation. From this point of view Maurey's apparently suicidal programme appeared in a different and much grimmer light, and reminded Sam that a hundred men who knew the basic uses of explosives could have taken over the Roman Empire by direct frontal attack.

Maurey had called the tournaments a blind for his moon colonization project. It was interesting to see what understanding could come out of standing Maurey's statements on their heads. Maurey had never had the faintest interest in the moon, as Sam knew he should have seen at once. On the other hand, it was now clear that the tournaments were essential – that they were nothing less than training grounds for a tetraploid militia, for training in a new and terrifying armamentarium.

But it was in the matter of the 'phony' tetraploidy that Maurey's massive intelligence had shone most brilliantly. Sam had a sickening hunch that there actually was something amiss in Sena's genetic background, but Maurey's moves made sense all down the line even if one assumed that the whole 'phony' story had been pure invention. The accusation had invalided Sam out of any participation in anti-diploid politics, a field where Maurey could not afford to trust him, first by giving Sam something more immediate to worry about, and second by providing the other giant with grounds for distrusting Sam.

The murder followed logically. It stowed Sam safely away physically, as the 'phony' story had already isolated him politically; and at the same time it multiplied anti-diploid feeling marvellously by making martyrs of both victim and accused. Finally, the records which had been taken from the safe had been selected with perfect cunning to suggest that the very tetras least likely to go along with Maurey were also 'phony' in their genetic make-up – and that Maurey himself might belong to that group, which made *him* look like an altruist.

It had been well done. Sam, lying full length on the cell bunk with his arms folded under his head and his feet on the cold floor, was surprised to find in himself an impersonal streak which found it all admirable. None of his deductions had thrown him into the expectable fury against Maurey. The renegade giant, in Sam's limited vocabulary, simply had no morals.

He was neither mad nor bad, but only a direct-actionist. He was a social outcast, like all the tetras, but he was unlike them in following exile into its last ditch – that murky declivity where there is no such thing as bad means. Maurey would not even excuse bad means by a good end, for Maurey would consider a bad end worth no means at all.

This trait in Maurey, Sam had seen often in the laboratory, usually with the result that Gordian knots fell asunder with magical suddenness. Early in the study of the one-way push, Sam had resisted a suggested line of inquiry on the grounds that it was mathematically ridiculous. His chief had said, 'Are you going or staying, Sam? Math is just rationalization after the deed. If you're going, *go*, and let the explanations wait; if you don't want to go, then stay home, but don't complain that the goal you've abandoned doesn't show on the map. If you decide to stay home, you don't care what the map shows. The place you *want* to go to always exists, even if it's marked *Terra Incognita* or "Here Are Dragons".'

Maurey was admirable. Nevertheless, living as a human being demanded a constant fight for protection against his logical kind. Sam had personal objections to mangling the lives of others for any end; and the same impersonality that allowed him to admire Maurey's clarity and brilliance made him ruthless against the ends Maurey sought. It would make him equally ruthless against Maurey himself, when the time came.

Sam had hardly a noticeable fraction of Maurey's intricacy, but he was sure that time was coming.

There were footsteps outside, and Sam propped himself up on one elbow. The guards had brought his dinner. By diploid standards they were tough and chunky animals, as formidable as bears – but then by diploid standards the bars of Sam's cell were impassable, while against Sam they had to be electrified.

The guards put Sam's tray on the floor before the door and backed off to either side, retrieving their shotguns. Both of them looked up, their faces lit by the bulb which, Sam had decided, showed whether the current was on or off; then the glow vanished from their stubbly chops and they looked down again. 'Come and get it, lummox.'

Sam got up and bent to pull the tray under the bars. As usual, the meal was heavy, more than double the ration for the biggest possible diploid, and so nearly double what Sam needed. Like all the giants, his katabolic rate was very slow, and a high proportion of what he ate served him as fuel rather than as material for the building of new cells. Evidently the prison authorities had assumed that he'd been returning half his meals uneaten because he'd been too nervous to clean the plates.

It was only one more sign that the people who had the best reasons to be concerned with the problem of the tetraploids had not made the smallest effort to learn the available facts about them, though all the facts had been available for half a century.

The guards watched, waiting for the light to come on again. They were stupid, but not unfriendly, despite the gingerly way in which they had to approach him. One of them said, 'Heard the news?'

'I saw the morning paper,' Sam said, denuding a chopbone. 'Something's come up since?'

'The gov'nor's put the kibosh on the big show you guys were going to stage,' the guard said. 'Says it might cause a riot. What was it goin' t' be like, anyhow? Was you rilly goin' t' fly through the air an' all that?'

'I wish I knew,' Sam said. 'I got clapped in here before I heard more than a rumour or two. Methfessel seems to have changed his plans since then.'

'It's a damn dirty trick, if you ask me,' the other guard said. The light came on, and they lowered their guns and came a little closer to the door. 'I bought tickets for the wife and kids – two bucks a throw for seats 'way up in th' bleachers. This Methfessel goin' t' give refunds or rainchecks?'

'Oh, sure,' Sam said. 'He'd have to. He's been running the sports for the university up to now – I'm pretty sure he's honest.'

'Well, the family's goin' t' be pitched off about missin' it, all the same.'

'You was lucky,' the first guard grumbled. 'They wasn't no tickets when *I* got t' the box-office. I got mine from a scalper at ten rocks a throw. One buck seats, too. I'm gonna lose nine apiece on 'em, an' a couple the boys is in the same fix. If I had this Methfessel I'd take it outa his hide – but I guess it ain't his fault, neither. It ain't as if I had money t' burn.'

'Tough,' Sam said sincerely. 'As far as I'm concerned, I don't think Methfessel had any business announcing the tournament to begin with. He should have known it'd just have made more trouble.'

'Yeah,' the guard said, rather automatically, since it had been obvious that he hadn't been listening. 'Looked like it was goin' to be good, too. Every onct in a wile we could see one of the big guys shootin' up above the stadium an' down agin like a freakin' eagle—'

'You could see—'

'Just accidental,' the guard said hastily. 'Not that we was peekin'. We paid our money fair an' square, so why should we of peeked?'

'Oh, I didn't mean that,' Sam said. But it was impossible to admit what he had meant by his interjection – he was almost afraid to think about it in public. His appetite extinguished suddenly, he put his tray back on the floor and slid it out. The guards, shrugging at his sudden reticence, took it up and went off.

Sam sat still on the narrow bunk, chill and stunned. So the talk about flying in Methfessel's ad hadn't been just hyperbole! Evidently Maurey, perhaps with help, had developed the one-way push into a sort of – well, a sort of boot-strap for self-lifting purposes. Of course it was now easy to see how such a thing could be designed – but Sam hadn't seen it before, all the same. Evidently he needed the Big Think much more than he'd been ready to believe.

Sam thought conscientiously.

After he was through thinking, he was still sitting on the bunk. This flat fact could not be thought away; it was the most important thing he had to think about. If Maurey had any sort of flying equipment – and even a 'flying belt' was not unthinkable – he would without doubt stage some sort of melodramatic rescue raid on the prison where Sam was held.

Sam found that pill hard to swallow, but he swallowed it. The subtleties of amoral persons, of 'expedient' politics, invariably wound up in just such cataclysmic crudities if they were pushed far enough along the line of their own logic, and Maurey was exactly the man to push that far. Maurey

was a genius in most respects, but his ability to boggle at the
verge of disaster was slightly below that of a lemming.

And no amount of thinking would turn up any more logical specific step than that of a raid on the jail. No other
percussion-cap for a showdown-by-force with the diploids
could be expected to crop up for some years longer than
Maurey's patience would last. All Maurey's plans pointed to
exactly that – indeed they were ultimately explicable only in
terms of that intent and no other. If Sam were freed, Maurey
would have to think of something else; but Sam did not
expect to be freed, nor acquitted – and to include the possibility of acquittal in his puzzling was to include nothing
more than a hope that problems would solve themselves.
Sam was too good a scientist to let that hope creep into his
hypothesizing.

FACT: Sam was *inside* the jail. Knowing Maurey's plans
with reasonable certainty did him no good whatsodamnever.

The raid could only precipitate a massacre. In the confusion Sam would perhaps get away, and afterward he
would have to be shown to the tetras who had freed him;
that meant that he would not be killed treacherously under
the guise of being done a favour. The chances were slightly
less good for Sena – who would probably disappear – not to
be killed at once, for if there were hidden in her any solution
to the tetraploid problem, however disagreeable, Maurey
would have to know it in its totality in order to combat it, as
that same knowledge would be needed to make it work; but
she would be taken out of circulation, and, eventually, elimi-
nated when Maurey became satisfied with what he knew.

In the face of all this, Sam knew himself to be nothing but
a hulking male Cassandra. He could, if he chose, tell the
truth of what was to come, but no one about would listen.
His actions were just as constrained as his words. He would
be tried; convicted; when Maurey's 'rescue' force arrived, he
would escape. To escape now, entirely aside from the fact
that it could not be done, would make him the object of a

merciless manhunt, which would turn into a pogrom before it was over – exactly as Maurey's raid would turn into a pogrom.

In both cases, there would remain some question as to who was supposed to be exterminating whom, until the very last poor dog was hung, and the remaining bloody noses counted.

FACT: Sam could not stop the raid. The situation had deteriorated to a point where the raid, no matter how mad it was bound to be, had to happen. No one could stop it, any more than the trial itself could be stopped. Event logic dictated it, and, too, dictated Sam's escape. After that—

After that, perhaps, Cassandra might step out of the play, in favour of Orestes returned from exile. *Maybe.* There was no better answer yet. For now, Sam had only one function in the drama:

He sat and waited.

VII

Sam's lawyer was young, short of stature, and implacably cheerful. His name was Wlodzmierzc, which is the kind of name newspapers never misspell (the Smiths are the unhappy cognomens which get ignored by the proof-reader). Wlodzmierzc was chatting with reporters now, as had been his practice during the past three days toward the end of every recess, switching effortlessly into one or another of six different languages as needed.

The additional languages were always needed, for the world press had taken up the story of the trial, and legal observers from the International Court of Justice also were present. Wlodzmierzc himself was a UN appointee who had presented his credentials to Maurey before Maurey had decided on a man in whom to invest the tetra's war-chest. Since the Pole was obviously better qualified than any possible lawyer Maurey could have hired, and since in addition none of the tetras could afford to give away money where there was no need, Maurey had been forced to pass back the sums to the original contributors.

Whether or not Maurey had been happy about this unexpected turn remained an open question. He had not confided in Sam. The imprisoned giant suspected, however, that Maurey had accepted it as inevitable, and therefore not worth more than a mild swearword or two.

Even Sam himself had seen it coming, as soon as the Soviet UN delegation, still smarting from having been forced to return five acquittals in the Belgrade Trials, had suggested the possibility of 'lynch law in the American giant case'.

Even the American representative had had to admit that 'some public prejudice might possibly affect the conduct of the trial'. After that, though the way to be traversed had still been tortuous, a Wlodzmierzc had been clearly visible at its end.

The bailiff rapped, and the lawyer came back quickly to the defence table, smiling innocently at Sam.

'Anything new?' the giant said in a low voice. The lawyer seated himslf and leaned sidewise; always, when seated, he watched the bench and the witness chair steadfastly, never looking at Sam, but canting alarmingly whenever they had to speak.

'Not very much, I am afraid. I am beginning to feel that international intervention here has had a largely ugly effect upon the local populace, and such an attitude inevitably will filter through to the jury and perhaps even to the judge. A pity that we failed to get a change of venue to England.'

'I wish you'd explain again why you tried that.'

'A question of publicity purely, Sam. English law does not permit newspapers or television commentators to discuss a criminal case until a decision has been handed down. Afterwards they may give complete accounts of the trial, and claim abuses of justice if they believe they have seen any, but beforehand, no. The UN's proposed World

Code includes such a provision, but the United States and several other—' He dropped abruptly into the telegraphic pidgin he used while the court was actually in session. 'No matter. Here's judge.'

'Sena Hyatt Carlin!' the bailiff cried.

The audience stirred. This was its first chance to look in person on the 'Blonde Princess' who had featured so largely in the tabloid accounts. Sena came to the stand confidently, was sworn in, turned and sat down with a concentrated grace. Her expression was clear, and a little cold – suggesting neither disgust nor contempt, but simply aloofness. Sam took a deep, quiet breath.

Both the aloofness and the confidence became more marked as she answered the preliminary questions. They came rapidly, and Sena answered them at the same pace, using the number of words proper to answer the questions as put, no more, no less, allowing the district attorney to establish her identity and her qualifications as a witness. In the same position Sam would have gone slowly, wary of possible traps in the first routine queries, but Sena did not appear to be afraid.

The prosecutor might have been impressed, if unwillingly, by her self-possession; in any event he set no traps. He said at last, 'Now, Miss Carlin, is it true that Dr Hyatt never informed you that you were not a tetraploid individual?'

'No.'

'No, he did not?'

'No, it is not true,' Sena said.

'Then he *did* so inform you?'

'No. He had no true information of that kind to give.'

The D.A. smiled. 'We'll let that pass for the moment. You have heard Dr St George's testimony – I refer to that part of his deposition in which a visit by him to the dormitory room of the accused is described. Is that description accurate to the best of your knowledge?'

'Quite accurate,' Sena said coldly.

'Did you, at the time of that conversation, believe that Dr St George might have been misleading you?'

'No, I didn't.'

'Very good,' the prosecutor said in succulent tones. 'You considered then, that there might be a real barrier, an impediment let us say, to your having children by the accused?'

For the first time Sena appeared to be slightly uncertain. 'I suppose I did feel that way,' she said at last. 'For the most part, though, I was just – well, alarmed, and anxious to find out whether or not M – Dr St George was right.'

'I will not protest that answer, but I must ask you in the future to confine yourself more closely to the question proper,' the D.A. said. 'Prior to Dr St George's disclosure, however, you had planned to marry the accused; is that correct?'

Wlodzmierzc snapped open like an automatic card table. 'Objection.'

The judge looked interestedly at the Pole, as one examines one's first fossil dinosaur egg. Wlodzmierzc said, 'My honourable opponent's question is so phrased as to open the question of the family system among the tetraploid people. Such material would be irrelevant and most certainly prejudicial.'

'The prejudicial aspects are clear,' the judge admitted, turning to the prosecutor. There was a slight edge on his voice, and Sam was instantly convinced that the jury had been meant to notice it. 'Mr Sturm, are you prepared to defend the relevancy of the material?'

'No, your honour; my phrasing was fortuitous. I will withdraw the question and restate it.'

Yeah, Sam thought glumly. *Now that the jury has been reminded of what a loose-living crowd the lummoxes are anyhow—*

'Yes, that's true,' Sena was saying.

'Thank you. What was the reaction of the accused to Dr St George's disclosure?'

'He didn't believe it,' Sena said.

'Quite; but Dr St George has also said that the accused was angry. Was that your impression as well?'

'No,' Sena said. 'Anyhow, not exactly. Do you mean whether or not he seemed angry at Dr Fred?'

The lawyer bowed ironically. 'That is what I meant.'

Sena shook her head. 'He was upset, just as I was, but he couldn't be mad at Dr Fred until he'd found out whether or not the story was true.'

'Then he *would* have been angry had the victim told him the story was true?'

'That would depend on the explanation. If Sam was shown a good reason for such a deception, I'm sure he'd go along.'

'This is, however, merely your estimate of the defendant's character.'

'That is what you asked me for, Mr Sturm.'

'True. When did you first learn of the murder, Miss Carlin?'

'That morning; I think it was about seven o'clock.'

Sturm smiled. 'One hour after the accused's appointment with Dr Hyatt, if I am not in error. And Ettinger notified you himself, I believe? Can you remember his exact words?'

'The first thing he said?'

'That will do nicely.'

'Yes,' Sena said. 'He said, "Check, Sena." '

The D.A.'s smile turned magically into a scowl. 'That's all?'

'Well, he said, "Good-bye," too.'

'Did you and Mr Ettinger play much chess?'

'No. I don't know how, and I've never heard him mention playing himself.'

'But I presume you knew what the accused meant by, "Check." '

'I thought I knew.'

'You're being rather stubborn, Miss Carlin. Must I ask you directly what your opinion is of the meaning of "Check, Sena"? Very well, what is your opinion?'

'He meant to ask me to check the genetic aspects of Dr St George's allegation. He saw, of course, that he was sure to be arrested and in no position to check the matter further himself.'

'Does it strike you that there are much simpler ways of interpreting the remark?'

Sam clutched Wlodzmierzc by the elbow, but the lawyer shook his head.

'Under the circumstances, no.'

'Then will you explain, please, how two enigmatic words would suffice to inform you of a murder, unless it had been foreplanned in your presence?'

There was a long-drawn *a-a-a-a-h* in the courtroom, general but too soft for any but a conscientious judge to silence.

Sena said, 'I already knew about the murder. Mr Ettinger sent for me and I saw the body before *any* words were spoken at all. After that not many words were needed.'

'But surely he spoke words over the telephone?' Sturm said gently.

'He didn't telephone; he sent a friend.'

'With what message?'

'No message.'

'The friend simply appeared? Who was this friend?'

'Dr Hyatt's dog.'

The attorney turned bright crimson in the space of a second. 'Miss Carlin,' he said in a tight voice, 'are you asking this court to believe that Mr Ettinger managed to get you to come to Dr Hyatt's laboratory merely by sending a dog after you? Or did he pin a tearful note to the dog's collar? Or was it, perhaps, a talking dog?'

'Which of your questions shall I answer?' Sena demanded angrily.

'None, Miss Carlin. None. I withdraw the questions. Mr

Wlodzmierzc, your witness.' The prosecutor made such a triumphal march to his table that Sam could almost hear strains of Meyerbeer in the stale air.

'One moment, Mr Wlodzmierzc,' the judge said nervously. 'You realize, I'm sure, that you may object to the final line of questioning before taking the witness, under American law? I am not suggesting that the prosecutor's questions were in any way improper, but I wish to be sure that you do not unknowingly forfeit any—'

'Thank you, your honour, but I have no objections to enter,' Wlodzmierzc said in a brisk voice. 'I am pleased that my worthy opponent brought up this question of the dog. Miss Carlin, rather than consume more court time bringing this information out piecemeal, I am going to make a brief statement about the dog myself; I shall then ask you whether or not the statement is correct, and if not, wherein it is in error.'

'I object!' Sturm said hotly. 'Your honour, surely the attorney of the accused is in no position to testify on behalf of a witness.'

'He has a clear right to pose a hypothetical question,' the judge said, 'depending upon its content, of course. Proceed, Mr Wlodzmierzc.'

'Thank you, your honour. Miss Carlin, this is my formulation:

'The dog in question is a giant dog. It is not a tetraploid, but closely related to the tetraploids, in a theoretical sense. As such, it is of abnormal intelligence, as well as of abnormal size. It was this dog, then, which awoke you shortly before seven on the morning of the murder – to use my learned friend's way of referring to a day during which no such murder may have occurred—'

'Objection!'

'Overruled,' the judge said unhappily.

'But your honour, the grand jury returned a true bill of murder!'

'Mr Wlodzmierzc didn't question that. He questioned the day.'

'—by entering your dormitory building, pushing your door open, and pulling the covers off your bed. We have testimony to show that this dog, this same dog, was seen and heard on the campus at this time, being very noisy; however, it made no sound while in the dormitory.'

'Are you prepared to substantiate this in any other way than by the passive agreement of the witness, Mr Wlodzmierzc?'

'Yes, your honour. We are prepared to bring the animal here, and to demonstrate that it can follow complex directions, understand situations involving as many as three variables, and exercise reasoning faculties in general which are slightly greater than those of a chimpanzee, particularly those faculties which might be termed integrative. I may go so far as to say that this dog is an important witness for the defence. In the meantime, however, I ask only that my statement be accepted as testimony from the *present* witness by virtue of whatever agreement she may vest in it.'

'All right. Let's hear the rest.'

'The rest is quickly told. Miss Carlin, you went with the dog to Dr Yatt's laboratory; she led you there. Once arrived, you found Mr Ettinger and the body. Mr Ettinger pointed to the spilled papers which have been mentioned before in the court, and said, "Check, Sena." You thereupon looked at all the papers during the next five minutes, and left that laboratory with the dog. At this point let me ask you whether or not I have stated the facts correctly.'

'Quite correctly, Mr Wlodzmierzc.'

'Good.' Wlodzmierzc darted with a sudden, sparrow-like movement to the defence table and returned, bearing a sheet of paper. 'Your honour, I have here a sheet of paper of ordinary legal length. It is completely covered, as these warranted duplicates will show, by single-spaced, typewritten lines, ungrouped, which consist entirely of figures. We prepared this document in hope of providing something which could not possibly be memorized in advance; Miss Carlin, in any event, has never seen it before. Will the prosecution allow us to show it to her for four seconds by stop-watch, in order to demonstrate that she is able to memorize it with complete accuracy in that time?'

'Well, Mr Sturm?'

There was a good deal of mumbling, during which Sam was vaguely surprised to find himself in the throes of a chill of malarial violence. At last Sturm agreed to let Wlodzmierzc proceed with the demonstration, providing that Sena would also memorize in four seconds two pages of the FAO rice-production tables for 1948, to be selected by Sturm.

Sena did beautifully with both, muffing (as Wlodzmierzc had before the trial insisted that she should) eight of the thousand figures in the prepared sheet, and throwing in a stumble over one word in a footnote on the FAO tables for good measure.

'We have arranged this demonstration, Your Honour,' the Pole said, 'in order to establish that Miss Carlin is capable of memorizing written information in great quantities, practically instantaneously. Miss Carlin, will you confirm?'

'I have what is often called an eidetic memory,' Sena said composedly.

'And, Miss Carlin, did you so memorize the contents of Dr Hyatt's papers while you were in the laboratory with the accused?'

'Yes, sir,' Sena said. 'There was plenty of time for that; I believe I went through them three times to make sure I had seen everything.'

'Your Honour, these papers are in evidence. If the court or the defence so wishes, Miss Carlin is prepared to quote at length from any given page as a further check.'

The judge looked at Sturm, who shook his head. Wlodzmierzc said: 'The reason why we have been at pains to establish this fact will appear in a moment. Now, Miss Carlin, I am going to ask you a very important question, and I want you to consider your answer most carefully. This is the question: Did you, or did you not, see anything in those papers relating to your presumed non-tetraploid status?'

'That's very easy, Mr Wlodzmierzc. I did.'

'According to what you saw, are you a tetraploid individual?'

'No, sir.'

The crowd murmured, but Wlodzmierzc was not through. 'Is the accused?'

'No, Sir.'

'Is Dr St George?'

'Objection!' Sturm said. 'Dr St George is not on trial. The question invades his right of privacy.'

'Sustained,' said the judge.

'Very well. Let me then ask you this, Miss Carlin: Of the entire colony of giants, how many, according to your information, are tetraploid individuals?'

'None,' Sena said flatly.

There was a roar of incredulous amazement in the court. The judge made no attempt to control it. After it had died down however, he said, 'Mr Wlodzmierzc, almost I suspect you of provoking that statement sheerly for confusion's sake.'

'Not guilty, Your Honour; the testimony is exactly relevant. Miss Carlin, does the accused know this fact – that is, had he known it up to this moment?'

'No, sir, not to my knowledge. I believe no one knew it but Dr Hyatt's personal assistants, and even among them it was customary to refer to us as "tetras".'

'Why was it?'

'Because it was convenient. Since every one of us has a different degree of polyploidy, and of a different kind, some over-all handle was needed. The only other choice would have been "Polly".'

'I see. What, in your opinion, is the source of the confusion?'

'In the use of the term "diploid" for people of "normal" genetic constitution. The "normal" human being actually is a *tetraploid* individual, like the tomato and certain other—'

This time the judge did pound for order, looking both baffled and wrathful.

'—but the doubling of the chromosomes apparently happened millennia ago, so that geneticists customarily speak of redoubled humans as tetras because they've twice the normal number of chromosomes. Actually, of course, such an individual would be an octoploid.' She smiled. 'Except that there are only two such individuals in our colony, we might have been nicknamed "octopuses", I suppose.'

'Thank you. Your witness, Mr Sturm.'

Sturm got up. He seemed considerably shaken, but he advanced grimly upon Sena. 'Miss Carlin, are you a geneticist?'

'No, sir.'

'Have you ever had any training in genetics?'

'I have had one two-semester course.'

'Did the late Dr Hyatt personally tell you any part of the hypothesis you have just offered the court?'

'No, sir, as I told Mr Wlodzmierzc.'

'Have you checked any part of this hypothesis with the personal assistants of Dr Hyatt whom you mentioned?'

'Briefly. Dr Edwards agrees with it. Dr Hammersmith was more cautious and said only that it might easily be true.'

'Did he state the reason for his caution?' Sturm asked drily. He was beginning to recover some of his composure.

'Yes, sir. He said that no one knows whether the "normal" human being is tetraploid or not; that it was probable but that it hadn't been proven. He did add, however, that he had often discussed the point with Dr Hyatt, and that Dr Hyatt maintained that his experiments with us, with the giants, were close to clinching it.'

'We'll ask Drs Edwards and Hammersmith to testify later. Will you state again whether or not the accused had any knowledge of this hypothesis?'

'I believe he did not,' Sena repeated.

Sturm nodded to the jury. 'Then it could not have affected his conduct on the day of the murder?'

'No, I don't see how it could have.'

'Now, about those papers. Were papers relating *directly* to you among them?'

'No, sir.'

'Or to the accused or to Dr St George?'

'No, sir.'

'Then you are unable to say exactly what your genetic status, Mr Ettinger's, or Dr St George's might be: is that correct?'

'Quite correct.'

Sturm straightened and said in a harsh voice, 'Your Honour, the prosecution feels that further pursuit of this aspect of the case would be fruitless. The prosecution rests its case.'

The judge looked at Sam's counsel. 'Mr Wlodzmierzc, has the defence any further witness to call?'

'Yes, Your Honour. We wish to bring the triploid dog Decibelle to the stand, demonstrate her intelligence by appropriate tests, and ask her certain questions of a nature which, as shown by the tests, she is capable of answering.'

Sturm shot back to his feet, gesticulating wildly, but the judge was ahead of him. 'Mr Wlodzmierzc,' he said in a gravelly voice, 'this is an American court of justice, not a sideshow or a music-hall. The court has permitted you to introduce certain facts concerning this dog, but neither human patience nor the dignity of the law can countenance introducing this animal as a witness. If you have any further *admissible* witnesses to call, please do so. If not, this court is in recess for today.'

The summations took all the next day, but the jury stayed out only six minutes.

* * *

VIII

The effect of the verdict upon the public temper was astonishing, especially to Sam, whose knowledge of late Roman history was about as extensive as that of any other layman – in short, zero.

Up to the first day of the trial, the question of whether or not Sam was guilty had not been much debated. It had been assumed generally that he was guilty. The actual guilty verdict, however, seemed to open up a wide gap in the populace; suddenly, the air was charged with dissension.

The letter columns of newspapers were filled with communications of terrific virulence of language, each writer denouncing a previous one, and/or the stand of the paper itself. Fights over the subject in bars, sometimes involving all the customers, the barkeeps, the entertainers and the cops who came to restore order, became outright commonplace.

Clergymen unlucky enough to announce 'Atlantean' opinions – which most of them held – in predominantly 'Titan' parishes lost their posts. Video commentators of opposing views raked each other recklessly over the coals. The sports pages of the papers teemed with cartoons and columns about the controversy. Senators made 'Titan' and 'Atlantean' speeches to their constituents while campaigning – sometimes gauging the prevailing opinions in their constituencies with great inaccuracy. Slanderous denunciations became too common to merit headlines any more and 'tetra' libel suits burst out with the frequency and violence of popcorn.

The whole complicated issue was further clouded by a heavy political coloration. For some reason, the general 'Titan' viewpoint was adopted by most of the left-wing elements of the population, all the way from mildly pro-labour groups to militant socialists; the conservatives, on the other

hand, espoused the 'Atlantean' point of view, which was not
only anti-Sam, but, unlike the team for which it was named,
was also anti-giants too. Embarrassingly enough, the rem-
nants of the American Communist party also adopted the
'Atlantean' creed, claiming the giants were laboratory
zombies created to further capitalist schemes of world domi-
nation.

This coloration carried the bitter quarrel all the way
into the home. Sons and daughters ordinarily took the
'progressive' Titan line, while their parents registered stiff
Atlantean disapproval. The subject was complex enough
to nurture family splits as rancorous and as final as the
theological hairsplitting which had been the bane of other
ages.

Most of these developments Sam had to deduce, not without amazement, from the papers brought to him in the
death-cell. The first riot, however, he saw from the window
of his own cell. A small labour union had arranged a 'Free
Sam Ettinger' demonstration just outside the prison, in response to the Intern the Lummoxes' campaign which one of
the yellower newspaper chains had been pushing. Similar
demonstrations had already been held elsewhere in the city,
all of them innocuous and, of course, ineffectual.

But this one was outside the prison. The governor, a Titan
himself, but ridden at home by an Atlantean faction, was in
a bad state of the jitters; he committed the tactical error of
calling out the militia against the demonstrators.

Most of the marchers were skilled workers in an engineering trade involving considerable training; they were peace-
able, intelligent men in their forties, who would no more
have stormed a prison than they would have taken to piracy.
The arrival of the state guard threw them into a state of high
indignation. Furthermore, a mob of Atlantean factionists
who had gathered to jeer at their Titan enemies got in the
way, were shoved aside, and promptly began to stone the
militiamen for interfering with the right of free assembly.

After that, Sam could not keep the two groups sorted out. There were shots, and tear gas, the men carried off in ambulances, and windows broken. The whole riot moved off from the prison, disappearing into the city proper, getting louder as it went; and inside the prison, a siren was howling – not because there was, or had been, the slightest chance of a jailbreak, but simply because the warden had been unable to think of anything else to do.

All of which, Sam knew, was only a prelude to holocaust. He went back to his bunk and waited for it to happen . . .

It began with a soft, hornet-like droning, not somnolent and soothing like the burr of bees, but with a harsh black edge on it, part hiss and part snarl.

Sam heard the angry midnight sawing for some time before it became distinct enough to be marked. He got up again and went back to the window.

He was aware that the sound had been going on for some time, but up to now he had not dissociated it from the rumble of the never-silent city. His heart and his breathing began to misbehave, and his mouth was very dry.

The future looked both short and violent from the black window. He had never, at any time, expected to be acquitted, but the court's refusal to allow Decibelle as a witness had killed the only real hope he and Wlodzmierzc had had – the hope of implicating Maurey sufficiently to impede him. Wlodzmierzc had known about the raid, but he had warned that only the dog could point definitely to Maurey as the real killer, and that Sam's own life hung from that probably inadmissible accusation.

That accusation – though neither Wlodzmierzc nor any other non-giant could know it – could not have saved Sam's life; but it might have determined how usefully he died. Once it had been made, in public, Maurey would not have dared to engineer any coup.

A small, black clot, granular, like coal dust, was gliding

out of the horizon along the dully-lit undersides of the clouds. The humming grew steadily. So did the clot.

Sam wondered desperately why the local army base had not already been alerted. Surely they had searchlights and anti-aircraft weapons there. And what was the matter with the Air Force's radar net? A few jets in the air now would make all the difference—

But no lights went up, there was no sound of planes; the city, exhausted by the riot, dulled by a vaguely soothing speech from the governor, snored. Belatedly, Sam realized that the humming sound was only just above the threshold of audibility – it sounded so enormously loud to him only because it had meaning for him. If only he had, after all, told Wlodzmierzc, the warden, the court, anyone, what he knew to be coming; someone, someone would have believed him, or have been alarmed enough to sleep badly, to straighten now in his bed and ask himself, *What's that?*

Perversely, now that he had conceived the hope of its being noticed, the humming dwindled in Sam's ears and blended back into the somnolent droning of the city. For long seconds at a time he was convinced that he could not hear it at all. Then – since it had not changed at all, except to come a little closer – it sprang back into being around his head like all the hornets of Hell's own ante-room.

The grains in the clot separated, became little black bacilli against the lurid culture-medium of the sky. The humming was now so heavy as to make Sam's eardrums flutter uncomfortably; he realized suddenly that it was too loud for the apparent disturbance of the swarm. Lights were coming on in the city, too, and somewhere deep in the prison there was a hoarse shout of alarm – not the shout of an official, but that of a trapped man whom doom approaches.

The humming swelled again, growing so suddenly almost to a roar that Sam ducked involuntarily. When he looked again, a swarm of clearly definable human figures, silhouetted inkily against the sky, was pouring over the prison

– was pouring *away* from him, toward that other cloud which had come from the horizon.

A thin spear of monochromatic yellow light stabbed from the clenched fist of one of the near-hurtling shadows. There was a flat crack, not nearly as sharp as the sound of a gun, but somehow reminiscent of thunder, all the same. Immediately, there was a fusillade of them.

The more distant, oncoming group responded at once. No sound could be heard from it, but the flying cloud was stippled with yellow stars. At the same instant, Sam's eyes were filled with stone-dust, and a fearful blow across his skull, just above the left temple, slammed him reeling away from the window.

In the darkness, his head ringing, his gritty eyelids burning, the bitter truth drove in upon him. There were 'Atlanteans' and 'Titans' among the giants, too. Maurey obviously had whipped up a predominantly Titan group to staging a raid on the prison – but the Atlanteans, in surprising strength, had gotten there first.

A pitched battle, a civil war in the air, was already under way – and not just between giants and diploids, but between giant and giant.

He stayed away from the window, his eyes watering. He had no idea of the power of the version of the one-way push which the flying squads were using as a weapon – the gaudy spears of light, he deduced, were stigmata of the adaptation of the principle to stadium use – but as its discoverer he knew already that it would be effective over any distance, limited only by the horizon. The accidental, random hit on the window, from some shot fired by the still-distant Titans, had given him a more than adequate reminder of that.

Raging, he blinked away the dust and watched the development of the New Pasadena from the far side of the cell, through an embrasure of the apparent size of a postage stamp. The noise of the city was up a little, a drone-bass for the stuttering implosions of the giants' side-arms. A sudden wavering rib of light, appearing and disappearing in the field of the postage stamp, told Sam that the airfield, at long last, had come awake, and was groping for the cause of the disturbance in the sky.

At once, a whole series of heavy impacts struck the near wall of the prison. The shouting of the invisible prisoner rose to a wail; then it was drowned out by the prison siren – apparently the siren was the warden's only answer to all problems. Another series of blows followed, battering the stone with a sound like the merciless hammering of age-split hollow logs.

The Titans were taking no chances. Now that the threat of discovery from below had become serious, they were not wasting shots upon their indistinct Atlantean brothers. They were bombarding the prison, an object which they had some chance of hitting. It was even possible that they had little to fear from the Atlantean attack except at very close quarters – if both sides were using tournament weapons, they probably had effective armour against those weapons; Kelland would have been careful about a thing like that. The stone walls of the prison, on the other hand, would soften in a hurry under a reactionless bombardment.

The sirens howled on, completely obliterating all sounds from outside. But the shocks against the outer wall could be felt.

Then the corridor light went out.

Sam spun and stared. A maddening square illusion, about the size of a postage stamp, floated in front of him wherever he looked. It took a long time to fade, but finally he was sure. The lights were really out – all out—

Even the warning light which showed that the bars of the door were electrified!

Someone had been sufficiently frightened by the bombardment to pull the master switch.

The electrically-operated locks which kept all the cells closed would still be in operation, of course, powered by an independent 'hurricane' generator. But the charging on the bars had just been a jury-rig from the main lines. Sam grappled for the bars, and after two swipes, one sweating palm closed around cold steel. No lethal shock convulsed his muscles.

Bracing himself, he began to pull.

The door was tough. It seemed immovable. Then, it gave, just a little. His hand slipped; he wiped it on his prison dungarees and took a fresh grip, this time with both hands.

He was not going to be 'rescued' by Maurey St George if he could help it.

He got to work on another bar, dragged it painfully in the same direction as the first. There was no chance that he could get two of them far enough apart to allow him to slip between them; they were too close-set, and bending one meant bending all, since they were all bound together in a two-dimensional sheaf by four cross-pieces. But if he could bulge the whole cagework enough to drag the lock down and out of its socket –

The siren died abruptly. But no lights came on, and no searing shock raced through the bars. Outside the crepitations of the giants' weapons came through loudly; and now, too, there was an occasional, heavy *blam*.

Anti-aircraft shells.

Sam pulled. The hinges ground against the stone. One corner of the door scraped protestingly against the concrete floor. Sam bent, seized that corner, and forced it out and towards the centre of the cell—

With a coarse screaming, the bolts sheared. The thousands of foot-pounds of drag testing their small cross-sectional strength had told. The door came inward – against its normal direction of movement – about seven inches. Sam crammed himself between it and the wall, shoving with all his strength—

And was out in the corridor.

Ten minutes and two killings later, Sam Ettinger, the gentlest of giants, was at large in the terror-capped city.

Kelland pulled an edge of the blind away from the window, with the delicate movements of a man who half expects the material to tear in his hand, and peered with one eye around it out at the dim, eventless woods. Then he sighed, let it fall to, and turned on one shaded lamp.

'We're on the spot, Sam,' he said heavily. 'I did my best to keep those tee-total damned fools from staging that raid, but I couldn't get a soul to listen to me. I'm just supposed to design weapons and play dumb about what's done with them. Anyhow, you were lucky to get away, and I'm glad to see you. Is there any hope for salvaging anything?

'I don't know,' Sam said carefully, easing his weary, burning feet out in front of him. He had run most of the way to Kelland's isolated, ramshackle house, after he had managed to fight clear of the panic around the prison and get out of the clogged centre of the city. 'There may be. I was hoping that you wouldn't be part of either party on the raid, but I wasn't sure. I'll confess that I came close to bursting into tears when you opened the door to me.'

'That's all right,' Kendall said, his own feet suddenly seeming two sizes larger. He shifted in his chair. 'Forget it. Where's Maurey; do you know?'

'No, don't you?' Sam said, astonished.

'No, Sam. He was supposed to lead the Titans to the prison, but he never turned up. They waited for him half an hour, and then someone came in shouting about an Atlantean counter-raid. They all took off in a complete rabblement. A fanatic named Briggs – I think you remember him, the tetra who did Methfessel's first propaganda work? – well, he took Maurey's place.'

Sam groaned. 'And here we sit, waiting to be arrested, while the giants help the diploids to destroy us! Kelland, you built all this apparatus; I don't know to what uses you modified my principle. Isn't there any step you can suggest?'

'Well,' Kelland said cautiously, 'I can at least find out how the fighting is going.' He got up and took down a golden helmet from a high bookshelf. 'You needn't worry too much, you know, Sam. That force of yours has polarity – don't look so flabbergasted, did you ever encounter a field that *didn't* show polarity? – and I took the trouble to make direct connections between the armour and the projectors. They can't do much more than stun each other—'

'All right, all right, but they'll be massacred by the diploids when they come down!' Sam shouted. 'Bullets don't carry charges to be repelled by like charges!'

Kelland looked alarmed and settled the helmet on his huge head.

'Briggs? Briggs! Ah; good. This is Kelland. Did you lose anybody to the anti-aircraft shelling? . . . Well, that's not as bad as it might have been. Good thing you had sense enough to get out of the air. Why don't you pull out? . . . You *have*? But great God, Briggs, there's no sense in that – Sam's escaped. Get out of that concrete tomb before somebody puts the lights on . . . Never mind the Atlanteans. They can hear me as well as you can. They know Sam's gone. Think about the future of the giants for once! Get out before the diploids trap you in there. They may decide to blow the whole place up, guards, prisoners, and all, just to trap you in there. . . . Dammit, Briggs, you're a fool, and a giant fool is a bigger fool than a little one. Get everybody out of there. They'll trace the plant here sooner or later, and if you're in the air then it'll be a long fall!'

Sam sat bolt upright. Kelland looked at him, raising his eyebrows resignedly, lifted the helmet and held it delicately in his hands. 'They're in the prison,' he said, 'fighting with the Atlanteans and the diploids, but he isn't sure about our own losses. I can't seem to get any sense into his head. He doesn't believe that you're out. You'd almost think he wanted to shoot you down himself, he's so eager to locate your cell.'

Sam let that pass. '*Can you cut off their power?*' he whispered.

'Why, sure,' Kelland said, turning the helmet reflectively. 'I didn't think it safe to power each suit individually. Neither did Methfessel. We wanted a way to ground both teams if tempers got lost and the tournament showed signs of turning serious.'

'Where's the generator?'

'Right here, under the house. They pick up the broadcast from the helmet transcaster – the little Christmas tree you see here.'

'Kelland,' Sam said grimly, 'give me that helmet!'

Puzzledly, Kelland handed it over, adjusted the cheekmike to Sam's face. Sam said: 'Briggs, you've got five minutes to get out of there.'

'Mind your own business,' a harsh voice snarled inside Sam's skull. 'You chose to stay home. We'll conduct our business without you – and remember you afterwards. Right now, keep your nose clean, or . . .'

Briggs' voice trailed off. When it came again, it was shockingly different. '*Ettinger, is that you?*'

'That's right,' Sam said calmly. 'Your boss has run out on you, Briggs. He's snitched Sena for himself, and left you all to massacre each other. What you all fail to do to each other, he thinks the diploids will finish.'

'You're lying!'

'Oh? Did you read the records of the trial? Don't you know why Maurey wants Sena? And why he wants the rest of us dead? But it's too late for you to begin thinking; I'd better do it for you. Get over here, to – the power source. Fast. All of you. That includes Atlanteans. Five minutes, remember; at the end of that time, I'm going to switch the whole fool lot of you right out of the sky.'

'You dirty butcher.'

'Not I,' Sam said cheerfully. 'I'm giving you a break you don't deserve. But be sure you're – over here – before the five minutes are up. After that, your pop-guns won't pop any more. *Git!*'

He took the helmet off. Kelland's eyes were bugging. 'Are you nuts?' Kelland said. 'They'll flay you alive for the threat alone. Half of them already think you're a traitor. And you didn't give me a chance to tell you – but every tetra wearing one of my suits heard every word you said. Are you trying to commit suicide along with the rest of us?'

'No. I wouldn't have bothered to say anything into your helmet if I'd thought I was talking only to Briggs. Anyhow, I'm not going to be here when they arrive, Kelland. I've got other business. I want Maurey.'

'I heard that part of it. I don't think I believe it, though.'

'I didn't believe it either, at first,' Sam said soberly. 'But it's true. He's the one who killed Dr Fred, not I. He found out, accidentally, that our polyploidy was very mixed in nature, and that the way its manifestations will occur in the generations to come is going to be mixed. It was Sena's records that drove the point home to him, and it hurt his sanity. He wanted the giants always to be *giants*, always obviously superior, always able to lord it over other humans. He planned to revenge Pasadena by running it in reverse – by wiping out the diploids with the weapon I gave him.

'But after the trial, he knew that could never happen. He knew that the future lay in the assimilation, and the gradual reappearance, of polyploid characteristics among "normal" people. So it seems that he's decided to ditch the obvious giants – trap all of them into destroying each other, with the happy collaboration of the diploids– while he, Dr Maurice St George, superman, sets himself up to become the father of the future.

'So now there's nothing left to him for his pains but Sena. As long as he can hide her, he can protect himself, he thinks, from the present, and be patriarch of all the generations to come. But he can't hide her, Kelland – because she's mine.'

'How'll you find her?' Kelland asked gently.

'Isn't Dr Fred's old vacation shack in these mountains somewhere?'

'I've never heard of it.'

'So you haven't, I forgot,' Sam said. 'Sena and I wouldn't have known about it either if Maurey hadn't opened Dr Fred's safe. There wasn't any other record of it, and Dr Fred himself never mentioned it. But it's around here, somewhere. And it's where Maurey had to go. He would hide himself among those papers; it's the only way he thinks. I should know.'

He stopped to think. 'Kelland, tell the diploid police about this; we'll need them. But not until our own people get here and you can make them understand what's at stake. Give me one of those helmets and I'll report back as I go, so you can all follow me. We'll have to smoke Maurey out ourselves, but we'll need the diploids to make it good. And we'll need to show them that we're acting in good faith.'

'You'd better take the whole suit,' Kelland said, even more gently than before. 'Maurey will be armed with it, too, and there'd be no point in being killed by your own discovery, when protection's available.'

'Okay.'

'Sam,' Kelland said. 'You haven't answered my question yet. Excuse me, I guess I haven't asked it yet.'

'What is it? Go ahead, Kelland.'

'How do you plan to find her – and him?'

'I have a friend,' Sam said, smiling suddenly despite his heavy breathing. 'Sena said you were keeping the friend here for me. Do you still have her here?'

Kelland looked stunned. Then his answering, delighted smile irradiated the room. 'Yes Sam,' he said. 'Your friend is here. Go ahead. I'll do as you say. Your friend is – hell, man, go to the door and call!'

Sam strode to the door and threw it open upon the frosty night. Behind him, Kelland added: 'And – good hunting, Sam!'

Sam called: '*Decibelle! Decibelle! Come to Sam! Come to Sam! Decibelle – here, here, to me! Decibelle, here, to me!*'

There was a glad and thunderous barking. Sam went out into the night.

IX

The forest was tar-black, and itchy with the small night-movements of an old woods in a populated and resorted area, the movements of creatures too small to tempt hunters and too adaptable to care where they lived, the movements of tattered leaves and scrub wood, mice, squirrels, sparrows dreaming, barn owls, roaches, moths, midges, all the noise of regrowth after too-heavy timbering.

Decibelle tugged. Sam knew approximately where Dr Fred's shack was, and, though he had never been there, probably could have found it in daylight by himself. But the urgent need now was to get to Maurey – before the inevitable marshalling of diploid justice against the insurgent giants set Maurey free.

And not only for the protection of the giants. For the protection of Sena, who was their future.

He stumbled over something which might have been a foot. It was hard to tell, because the pressor-field disturbed his footing. Trying to recover, he slammed face-first into a tree-trunk.

The golden casque clanged. Dazedly, Sam righted himself, holding the dog back with difficulty. On his back, the heavy flying apparatus hung uselessly. The pressor-field was useless for collisions of this sort, cushioning the blow not at all, since all its pressure was out and away from Sam, none of it

back to him. It could shove branches out of his way, and
protect him from mosquitoes, but it could not push over a
tree.

So much for invincible weapons, Sam thought. *They lack
discrimination.*

The pull along the rawhide that led to Decibelle continued,
and Sam let it draw him. That impact on the casque had been
rather heavy. Better see if the radio was still alive. He pulled
the cheek-mike over. 'Kelland?'

'Present,' the earphones said at once, amid the forest
murmurs.

'Good. Kind of thick going here. Kelland, do you have
any foreign language?'

'Uhm. Will French do?'

'No,' Sam said. 'All I know about French is that it forms
plurals with soundless X's sometimes. Besides, Maurey
speaks it, and I don't care to have him know what I'm up to.
You don't by any chance speak German?'

'*Doch gewiss*,' Kelland said, shifting immediately into that
language. 'My family name was Keller, until I changed it to
keep the tetraploid stigma off my relatives. But Maurey—?'

'No,' Sam said. 'German is a chemist's language, not an
engineer's – Beilstein and all that. I owe my own knowledge
of it chiefly to a boyhood enthusiasm for Wagnerian opera.
The enthusiasm didn't stick, but the language did. Now,
what's going on back there?'

'Nothing yet. I'm still waiting for someone to show up,'
Kelland said. 'Any further instructions?'

Sam stumbled and swore. Then he said, 'Well, give them
the whole story, and try to keep them in hand until I find Dr
Fred's place. I'll call you all over here after I see how the
land lies. Leave behind a cadre to defend your power source,
or we'll all be *kaput*. And, oh, yes, as soon as you can, send
off a couple of our biggest boys to kidnap somebody in
authority – an Army officer, preferably, or a state cop. Equip
the prisoner with a helmet, so he can hear what we're saying

and follow whatever develops, but don't of course give him any flying equipment or pistol. Bring him along when I yell for help. Got all that?'

'Um,' said Kelland, in the voice of a man who is taking written notes. '*Erzähl' Geschichte ... stehl' Sicherheitsdiener* (that covers all categories, doesn't it, Sam?) ... *gib' Helm ... stell' Hindtreffen.* That about it?'

'*Jawohl*,' Sam said, and then, '*Ouch!* I'm going to shut up for a while. Decibelle seems to be drawing me through a jungle-gym made of barbed wire.'

'All right. Hello, here comes Hammy Saunders. I'll sign off too, since I'll have to speak English for a while to explain things to the newcomers. But I'll keep the earphones on.'

Sam pushed the mike away again and said, 'Hey-Decibelle?'

Rrrrrghrrph.

'All right. Go ahead. I'm with you.'

Decibelle obviously was not tracing out any path the brittle Dr Fred could have taken to his lodge by custom. These mountains were ancient and much worn-down, so much so that visitors from the Far West usually affected not to see them at all, but even so Dr Fred would have needed an open path, without any steep grades, winding around the bases of the hills.

The dog, on the other hand, at the moment was dragging Sam sidewise out of a creek-bed full of brambles, and on up the side of a crumbly rock shelf. Obviously, she was taking him along the way she herself went to the lodge, when unencumbered by human company. It was both a compliment to Sam and one more evidence of Decibelle's ability to assess an urgent situation.

Sliding up the hill with all his aching tendons creaking, Sam was able to notice that there was a little of dawn on the sky. Not enough of it had leaked through the trees, down below, to have given him fair warning.

The dog was pointing. She had never been trained to

point, but evidently she had seen a trained dog assume the odd position and had figured out what it was for. Her mimicry would have won her no prizes at any show, for her hackles were up and her ears laid back, but Sam was hardly inclined to be a purist about it. 'Good girl,' he said softly: 'It's on the other side of the crest, eh?'

Decibelle's tail wagged once and straightened into the 'point' position again.

'Good. Lie down, Decibelle. Stay here. I'll be back.'

Sam scanned the dim hill carefully and then lay down, working himself up the slope on his belly. At the top, he cocked one eye cautiously over the edge.

The other side of the hill sloped more steeply, and the shack clung to it, looking out over a placid and lovely valley with a stream at the bottom. This slope was grassy; and at a radius of twenty yards from the shack, the grass bent away in a perfect circle. A pressor-field.

Sam considered the problem. It was about what he had expected to find. He knew, as well as anyone knew, the characteristics of the reactionless effect. Since the field worked only in one direction, it could be used as a shield, but *not* as a detector. Force exerted against it had no effect upon its generator; the field did not push back, either physically or electrically. That meant that the only road to perfect security for Maurey was an unremitting, twenty-four optical watch, either in person or by a video watchdog. If Sam kept out of sight, Maurey would have no way of knowing whether or not anyone was in the vicinity.

Maurey, of course, had heard the German conversations, but had no way of knowing what they meant. He might have guessed that he was in danger of being smoked out; on the other hand, the fact he had chosen Dr Fred's lodge as a hideout – and the existence of the pressor screen showed that he had – was a clear indication that he did not expect anyone to suspect the place. The screen was routine; it would baffle diploids, but Maurey could not have expected it to baffle the giants for long.

He simply had not expected Sam to figure out where he was hiding. In that he had overestimated himself and – for the last time – seriously underestimated his former assistant.

Sam withdrew a few feet and called Kelland.

'We're all here, Sam,' Kelland's voice said. 'And we've got our Authority. We had a little trouble with Briggs, but he lacks Maurey's deviousness. He tried to sell the rest of us on Maurey's programme by talking about the necessity of Dr Fred's death. It made convincing the rest of us very easy; we have Briggs salted away.'

'You can come on over,' Sam said. 'The place is on the side of a hill overlooking the valley on the far side of the old deer preserve. If you come in on it from the north, he won't be able to see you.'

'Sam!'

The voice was Maurey's. It sounded cool and amused. 'I've been listening to you. Whatever in the world made you think I couldn't speak German?'

'All right, let's hear you speak some,' Sam said.

'Don't be ridiculous. You've made a big enough fool of yourself already; you don't even recognize your friends. They've freed you, and you can't think of anything better to do than hatch infantile plots with poor Kelland, who can be convinced of anything.'

'Where are you?' Sam said.

'I'll tell you that when I can depend upon your common sense, not before. I'm not going to endanger the whole project for one man who doesn't know when he's well off.'

'Where's Sena?'

'She's right here with the rest of us. If you want any part in the world of the giants, Sam, you'd better have some sober second thoughts. Our patience is about worn out; in a little while we'll have to go ahead without you – and I don't suppose the little obsolete folk will deal kindly with you.'

A shadow drifted in front of Sam, and Hammy Saunders landed lightly beside him. Sam said, 'You could be right. It's happened before.' He pushed away the mike again. Maurey's pathetic fictions continued to purr in his ears.

'Hammy, he's got the place surrounded by a field. We'll have to undercut the rock below. Send three or four men down into the valley, under cover, and get to work on the hill with pistols, just below the effective limit of the screen. Don't start till I say "when".'

'Right.'

Hammy melted away. Sam sat on the hillside next to the dog and watched the dawn-colours brighten, pulling sweet-clover and sucking the nectar from the tight white flower-clusters. Maurey seemed to have signed off, at least for the moment.

Kelland and two other giants came in silently, dragging with them a frightened diploid in civilian clothing, ridiculous in outsize golden helm. Sam took one look and whistled.

The man was the governor of the state.

'*Sicherheitsdiener* covers a lot of ground, all right,' Sam said, amused in spite of himself. 'Sir, I'm sorry for this forc-ible abduction, but believe me, we mean you no harm. We mean no diploid person any harm. We're here to smoke out the one giant among us who's created the trouble, beginning with the murder of Dr Hyatt. We've been forced to bring you along as a witness to our intentions.'

The governor was grey with terror, but he had an inherent dignity which stood him in good stead. 'I'm forced in my turn to accept that, for the moment,' he said stiffly. 'I'll watch and listen, since I'm powerless to do otherwise. But you may as well know that I don't believe you.'

'There's no necessity for that. If you watch and listen with an open mind our case will prove itself. You've already heard my conversation with Dr St George. He's over the hill, in a shack that used to be Dr Hyatt's summer lodge. He's holding one of us with him, the girl about whom you've heard so much; Miss Carlin. He doesn't know yet that we're anywhere in his vicinity. When we turn him out, you may hear more than enough to give you the full story; at least, that's our hope.'

'There are militia scouring the whole countryside for you!'

'We know that; that's why we're here. Had the militia arrived before us, had we told the militia where Dr St George was and why we wanted him, there would have been a number of deaths – including Miss Carlin's. He's well equipped to stand off any normal siege, except one conducted by heavy artillery, which would destroy him and Miss Carlin without proving a damn thing. We mean to convict him out of his own mouth, with no loss of life. Doesn't that strike you as a preferable way of going about it?'

The governor passed a hand heavily across his forehead. He was swaying a little. 'Perhaps so,' he said. 'If there's any truth in it at all. I'm in no position to be judicious. I've been kidnapped by agents of a convicted murderer, Mr Ettinger. My view of whatever it is that you plan will have to be coloured by that situation. Go ahead. I'll pay close attention; that's all I can promise.'

'That's quite enough,' Sam said gravely. 'Your recognition of your bias is also the assurance I need that you'll not be swayed by it. Kelland, see if you can find a vantage-point for the governor that'll be out of the way of any possible fireworks, but will still allow him to see everything that happens. He's the most important person in this show, far and away, and ought to be guarded accordingly.'

'Right,' Kelland said. 'Governor, we'll have to haul you through the air again – for the last time, I hope.'

'So,' said the governor, 'do I.'

They went away, Kelland leading, followed by the two giants with the small ineffectual figure of the governor between them, skimming the tree-tops in a wide arc towards the opposite side of the valley. The sun was coming up on the left, reaching gradually into the valley itself; a great many birds were making a musical bustle. There was actually a small curl of woodsmoke coming up over the brow of the hill from the shack, blatant, self-confident, innocuous.

After a little while Kelland came back. 'The boys are working on the hill under the lodge,' he said. 'It seems to be mostly soft marl. No alarm from Maurey, nor any sign of suspicion at all, yet. If everything goes right, the shack ought to begin to totter in about five minutes.'

'Good,' Sam said. He got to his feet and began to climb, unhurriedly, his face calm, his fists clenched. He went over the brow of the hill and stood looking down on the lodge.

'Maurey,' he said. 'You've been found. We'll give you ten minutes to come out of there.'

Maurey began to laugh. 'Out of where?' he said. 'Sam, what a child you are! Did you think you could get away with that "I know where you're hiding" line on me? If you want to know where we are, I'll tell you; but not before I'm sure you won't sell us all out to the diploids.'

'There are no human diploids,' Sam said patiently, 'and you've been found. Come to the window, Maurey, and look up the hill.'

There was quite a long silence.

'I see,' Maurey's voice said at last. 'Well, I suppose no refuge is perfect. And I suppose you've a great howling mob of diploids on your trail. You'd best send them back, Sam, before they get hurt. Don't imagine that a shack on a hillside is all that remains of tetraploid power.'

'I don't. I don't imagine that you've any great army of giants in there, either, Maurey.'

Maurey chuckled again. 'I won't argue with you. I retain some shreds of respect for you, Sam; and I recommend most strongly that you pull out before the final battle breaks. It's all over for the diploids; nothing that you can do can change that. Why get hurt?'

'There are no diploids,' Sam repeated. 'Where is Sena?'

'Sena?' Maurey said. 'Why, here, with the rest of us.'

'I'd like to talk to her.'

'She's busy.'

Sam moved one hand. The hillside, the ledges of the valley, the hillocks, the grasses uttered giants; they stood everywhere, motionless, like the dragon's-teeth soldiers of Cadmus.

'Here are the giants, Maurey,' Sam said. 'You can see them, if you'll look. There are only two or three missing, at the most – not counting those that were caught or killed in your raid on my prison. One of the missing is Sena. Where is she?'

'She's here.' Maurey's voice was as confident as before; there was nothing in it to indicate that a bookcase-full of his own lies had just fallen forward upon him.

'Let her out.'

'She doesn't want to come out. She's got more sense than all of the rest of you put together. I don't know how the hell you sold your brothers on this stupidity, Sam – I suppose Briggs got killed. I can think of no other explanation. Killed trying to rescue *you*, Sam. Anyhow, there's no essential change in the situation. If all the rest of you have sold out to the diploids, Sena and I will work out the proper destiny of the giants without you. Go home and rot, all of you!'

'There are no diploids,' Sam said. 'Let us talk to Sena.'

Maurey was silent. After a while it seemed that he was not going to speak again; then, startlingly, his voice came in, loudly, urgently. 'The rest of you,' he said. 'Listen to me. You're committing suicide. You had power over the diploids in your hands, and you've given it over to the man who murdered your creator. I gave you a cause; I gave you the means to be free of the pigmies, truly and finally free. Are you going to give all that up now?'

'Briggs said *you* killed Dr Fred,' a voice that Sam did not know said. There was no telling which of the gigantic gilded statues had spoken.

'What does that matter?' Maurey demanded. 'Be realistic! I didn't kill Dr Fred; obviously Sam did; but Dr Fred's death was quite necessary. It provided us with the chance we needed to arouse the diploids against us. Dr Fred preached peace with the pigmies. We all know that no peace was possible. What was needed, what's needed now, is war. You have the instruments for that war in your hands, I gave them to you, and intelligently used they're invincible. And you have the occasion. You could sweep the planet.'

'You divided us,' another anonymous voice said. 'You made us fight each other.'

'Harmlessly,' Maurey scoffed. 'You can't hurt each other with the force-pistols. I saw to that. Naturally I saw to it that you quarrelled with each other – there was no other way to disguise the raising of any army of giants. But your weapons are deadly only against diploids.'

A whisper, eerie and disembodied, came from among the statues. '*How about the anti-aircraft guns, Maurey?*'

'How about them? Your losses were tiny. Sam himself admits that.'

There was a low rumble from the giants.

'Too bad, Maurey,' Sam said, implacably. 'The truth is out, you see. It came out at the trial. *There are no diploids.* All human beings are tetraploid. We – the giants – are polyploid, but we're all polyploid in different degrees. As giants we won't survive; but we can survive through Sena and the others like her, because Sena's children will look normal. They'll be able to blend back into society, and they'll allow their grandchildren to forget their heritage. Eventually the polyploid characteristics will begin to reappear, piecemeal, until the whole race is heavily polyploid, and then giants will be commonplace and not the subject of pogroms.

'But as for you, Maurey; *you are a Pasadentist.* A subtle

one, but a Pasadentist all the same. You found out about Sena, and you killed Dr Fred to keep that a secret. You pitted us against each other, in the hope that the normals would destroy us all while we were snarling at each other. Pinning the killing of Dr Fred on me gave the normals reason to hate us, and staging the raid on the prison gave them reason to wipe us all out—

'While you hid here, with Sena, planning to become the sole father of the polyploid humanity of tomorrow – sole father of the tough, long-lived race that will be needed to reach the stars.

'It was a good gamble, Maurey. But since it was insane, it failed.'

Maurey said, 'Ridiculous.'

'Then let us speak to Sena. If she's free and on your side, she has a helmet and has been hearing everything that was said. Let her speak.'

'Certainly,' Maurey said, calmly. 'As I remarked before, she's busy. I'll see if she wants to talk to you. Hang on.'

Silence. The sunlight was now almost down to the bottom of the valley, where Kelland's sappers were worrying the foundations of Dr Fred's lodge. The clouds were pink with innocence.

'Here she is.'

'Sam?' Sena's voice said quietly.

'Yes . . . yes, Sena.'

'I'm all right. There's no reason for you to worry about me at all. He's got a gun on me *but he won't dare shoot*—'

On the last word her voice faded abruptly, and Sam's earphones shivered. His answering gasp was half fright and half sheer admiration. She said so much in so few words—

'Speak up, Maurey!' he shouted. 'Any last lies?'

'Keep your distance, all of you,' Maurey said. His voice was tight, frigid. 'You're a pack of fools. Just remember that you're outside, and I'm inside with Sena. Sam's quite right!

Sena's the key; if the giants are to survive at all, it has to be through her. And if any one of you makes a move towards the ledge, she'll die. She was right, too! I didn't kill her for squawking; I only removed her helmet; I don't kill for little things; I kill for reasons only. Like Dr Fred. Go away. Go away. Your future is in my hands, and there's nothing you can do about it!'

Suddenly, the lodge sagged sharply to the right. A mass of rubble went thundering and foaming down the hillside into the valley. Maurey's incredulous scream made Sam's ears ring.

The flat-lying circle of grass stood up.

The failure of the foundations had cut Maurey's power-line.

'*Decibelle!* Get him, Decibelle! Quick Decibelle *get Maurey!*'

The immense animal charged towards the hut, impossibly low to the ground. A force-pulse shot out from the sliding structure, but it was high.

Decibelle launched herself and went through the small pantry window in an explosion of glass. Maurey screamed again. Sam found himself running. Another bolt from inside the lodge blew planking outward in splinters. Still another never got outside at all, but the chimney tottered and collapsed, dumping hot bricks on the roof.

'*Take her off. Take her off! I'll kill you all, I'll kill you —*'

Maurey had his dog at last.

Before Sam was much more than half-way to the lodge, the hillside beneath it gave way completely, and Maurey, logs, bricks, fire, Sena, future, past, dog, plaster, dirt, concrete, pipes, wires, life, liberty, pursuit and happiness slid in one big untidy chaos towards the pineneedle floor of the valley, just ahead of the sun's fingertips.

It took a little while to separate Sena's just-living body from the wreckage.

It took longer to separate Decibelle from the ruins of Maurey's throat.

But after a while the giants were gone, and the shaken governor too; and the sunlight touched the valley, all the way down to where the wreckage lay, and began to climb back up the slopes.